OFFICIAL SQA PAST PAPERS WITH ANSWERS

ADVANCED HIGHER

CHEMISTRY
2007-2011

Publisher's Note

We are delighted to bring you the 2011 Past Papers and you will see that we have changed the format from previous editions. As part of our environmental awareness strategy, we have attempted to make these new editions as sustainable as possible.

To do this, we have printed on white paper and bound the answer sections into the book. This not only allows us to use significantly less paper but we are also, for the first time, able to source all the materials from sustainable sources.

We hope you like the new editions and by purchasing this product, you are not only supporting an independent Scottish publishing company but you are also, in the International Year of Forests, not contributing to the destruction of the world's forests.

Thank you for your support and please see the following websites for more information to support the above statement –

www.fsc-uk.org

www.loveforests.com

© Scottish Qualifications Authority

All rights reserved. Copying prohibited. No part of this publication may be reproduced, stored in a retrieval system, or transmitted in any form or by any means, electronic, mechanical, photocopying, recording or otherwise.

First exam published in 2007.
Published by Bright Red Publishing Ltd, 6 Stafford Street, Edinburgh EH3 7AU
tel: 0131 220 5804 fax: 0131 220 6710 info@brightredpublishing.co.uk www.brightredpublishing.co.uk

ISBN 978-1-84948-229-5

A CIP Catalogue record for this book is available from the British Library.

Bright Red Publishing is grateful to the copyright holders, as credited on the final page of the Question Section, for permission to use their material. Every effort has been made to trace the copyright holders and to obtain their permission for the use of copyright material. Bright Red Publishing will be happy to receive information allowing us to rectify any error or omission in future editions.

[BLANK PAGE]

X012/701

NATIONAL
QUALIFICATIONS
2007

TUESDAY, 29 MAY
9.00 AM – 11.30 AM

CHEMISTRY
ADVANCED HIGHER

Reference may be made to the Chemistry Higher and Advanced Higher Data Booklet (2007 edition).

SECTION A — 40 marks

Instructions for completion of **SECTION A** are given on page two.

For this section of the examination you must use an **HB pencil**.

SECTION B — 60 marks

All questions should be attempted.

Answers must be written clearly and legibly in ink.

SCOTTISH
QUALIFICATIONS
AUTHORITY

SECTION A

Read carefully

1 Check that the answer sheet provided is for **Chemistry Advanced Higher (Section A)**.

2 For this section of the examination you must use an **HB pencil** and, where necessary, an eraser.

3 Check that the answer sheet you have been given has **your name**, **date of birth**, **SCN** (Scottish Candidate Number) and **Centre Name** printed on it.

   Do not change any of these details.

4 If any of this information is wrong, tell the Invigilator immediately.

5 If this information is correct, **print** your name and seat number in the boxes provided.

6 The answer to each question is **either** A, B, C or D. Decide what your answer is, then, using your pencil, put a horizontal line in the space provided (see sample question below).

7 There is **only one correct** answer to each question.

8 Any rough working should be done on the question paper or the rough working sheet, **not** on your answer sheet.

9 At the end of the exam, put the **answer sheet for Section A inside the front cover of your answer book**.

Sample Question

To show that the ink in a ball-pen consists of a mixture of dyes, the method of separation would be

> A   chromatography
>
> B   fractional distillation
>
> C   fractional crystallisation
>
> D   filtration.

The correct answer is **A**—chromatography. The answer **A** has been clearly marked in **pencil** with a horizontal line (see below).

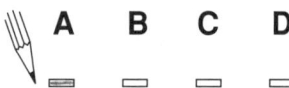

Changing an answer

If you decide to change your answer, carefully erase your first answer and using your pencil, fill in the answer you want. The answer below has been changed to **D**.

1. What type of bonding exists in an element which forms a gaseous oxide?

   A   Ionic

   B   Covalent (polar)

   C   Covalent (non-polar)

   D   Metallic

2. Which of the following compounds is likely to have the **most** covalent character?

   A   Tin(IV) iodide

   B   Iron(II) chloride

   C   Lithium fluoride

   D   Potassium bromide

3. Which of the following is **not** a form of electromagnetic radiation?

   A   Beta radiation

   B   Gamma radiation

   C   Ultra-violet radiation

   D   Infra-red radiation

4. **All** noble gases are characterised in terms of electrons by the completion of the outermost orbital. This orbital is

   A   an s-orbital

   B   a p-orbital

   C   a d-orbital

   D   an s or p-orbital.

5. Infra-red radiation can be used in the analysis and identification of organic compounds. Compared to visible radiation, infra-red radiation has a

   A   shorter wavelength and higher frequency

   B   longer wavelength and lower velocity

   C   longer wavelength and lower frequency

   D   shorter wavelength and higher velocity.

6. When an ammonia molecule accepts a proton to form an ammonium ion, there is a change of shape from

   A   pyramidal to tetrahedral

   B   pyramidal to trigonal planar

   C   trigonal planar to pyramidal

   D   trigonal planar to tetrahedral.

7. Isoelectronic species are molecules or polyatomic ions with the same total number of electrons.

   Which of the following contains two isoelectronic species?

   A NH_4^+ and CH_4

   B NH_4^+ and BF_3

   C BF_3 and NH_3

   D PF_5 and BF_4^-

8. A superconductor is a material

   A   whose electrical conductivity decreases with decreasing temperature

   B   that does not conduct electricity unless doped with another material

   C   that can conduct electricity with zero resistance

   D   whose electrical conductivity increases with increasing temperature.

9. The statement that "an orbital can accommodate, at most, two electrons, and if so, they must be of opposite spin" is based on

   A   Hund's rule

   B   the aufbau principle

   C   the Pauli exclusion principle

   D   Heisenberg's uncertainty principle.

[Turn over

10.

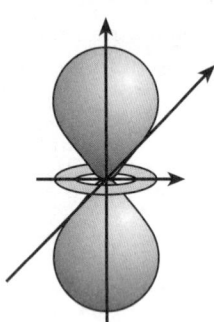

The above is a diagrammatic representation of the shape of

A   any p-orbital

B   a specific p-orbital

C   any d-orbital

D   a specific d-orbital.

11.   An ionic compound is thought to contain hydride ions. One way to show this could be to

A   electrolyse the molten compound and test for hydrogen gas at the negative electrode

B   electrolyse the molten compound and test for hydrogen gas at the positive electrode

C   dissolve the compound in water and test for an acidic pH

D   dissolve the compound in water and test for an alkaline pH.

12.

X   Y

The number of atoms of gas in flask Y is approximately

A   equal to the number of atoms of gas in flask X

B   twice the number of atoms of gas in flask X

C   one half the number of atoms of gas in flask X

D   one quarter the number of atoms of gas in flask X.

13.   In a gravimetric analysis of silver, a precipitate of silver(I) chromate was produced by adding excess potassium chromate to a solution containing silver(I) ions.

If $5 \cdot 795$ g of Ag_2CrO_4 was produced, the mass of silver in the solution was

A $1 \cdot 884$ g

B $3 \cdot 318$ g

C $3 \cdot 769$ g

D $8 \cdot 910$ g.

14. $N_2O_4(g) \rightleftharpoons 2NO_2(g)$ $\Delta H^\circ = +57 \, kJ \, mol^{-1}$

Which of the following will increase the equilibrium constant for the reaction?

A   Use of a catalyst

B   Increase of pressure

C   Increase of temperature

D   Decrease of temperature

15.   An aqueous solution of an organic acid, **X**, was shaken with ethoxyethane until the following equilibrium was established.

X(water) $\rightleftharpoons$ **X**(ethoxyethane)

X in ethoxyethane

X in water

$20 \cdot 0$ cm^3 of the upper layer needed $15 \cdot 0$ cm^3 of $0 \cdot 010$ mol l^{-1} NaOH(aq) for neutralisation.

$20 \cdot 0$ cm^3 of the lower layer needed $12 \cdot 0$ cm^3 of $0 \cdot 010$ mol l^{-1} NaOH(aq) for neutralisation.

The value of the partition coefficient is

A $0 \cdot 80$

B $1 \cdot 25$

C $1 \cdot 33$

D $1 \cdot 67$.

16.

$2Fe(s) + \frac{3}{2}O_2(g) \rightarrow Fe_2O_3(s)$ $\Delta H° = -822 \, kJ \, mol^{-1}$
$C(s) + O_2(g) \rightarrow CO_2(g)$ $\Delta H° = -394 \, kJ \, mol^{-1}$

Which of the following is the standard enthalpy change, in $kJ \, mol^{-1}$, for the reaction shown below?

$Fe_2O_3(s) + \frac{3}{2}C(s) \rightarrow 2Fe(s) + \frac{3}{2}CO_2(g)$

A   +231

B   +428

C   +1216

D   +1413

17. Which of the following can **not** be determined by a single experiment? The enthalpy of

A   formation of ethyne

B   combustion of methanol

C   solution of sodium hydroxide

D   neutralisation of hydrochloric acid by sodium hydroxide.

18. Consider the following bond enthalpies.

Bond	Enthalpy/kJ mol^{-1}
Br — Br	194
H — Br	362
C — H	414
C — Br	285

What is the enthalpy change, in $kJ \, mol^{-1}$, for the following reaction?

A   −1255

B   −39

C   +39

D   +1255

19. Which of the following is exothermic?

A $K(s) \rightarrow K(g)$

B $K(g) \rightarrow K^+(g) + e^-$

C $\frac{1}{2}Br_2(\ell) \rightarrow Br(g)$

D $Br(g) + e^- \rightarrow Br^-(g)$

20. Which of the following is likely to have the lowest standard entropy value at 100 °C?

A   Neon

B   Mercury

C   Sulphur

D   Phosphorus

21. An Ellingham diagram shows how $\Delta G°$ for a chemical reaction varies with

A $\Delta H°$

B $\Delta S°$

C   time

D   temperature.

22. Which of the following reactions must be exothermic? One in which

A $\Delta G°$ is negative

B $\Delta S°$ is positive

C   both $\Delta G°$ and $\Delta S°$ are negative

D   both $\Delta G°$ and $\Delta S°$ are positive.

23. In the electrochemical cell

$Cr(s) \, Cr^{3+}(aq) \, Ag^+(aq) \, Ag(s)$

operating under standard conditions, which of the following is **not** true?

A   The emf is 1·54 V.

B   The silver electrode decreases in mass.

C   Oxidation takes place at the chromium electrode.

D   Electrons flow from chromium to silver in the external circuit.

[Turn over

24. The following reaction is first order with respect to each of the reactants.

$$A + B \rightarrow C + D$$

Which of the following is correct?

A   The rate of the reaction is independent of the concentration of either A or B.

B   The overall reaction is first order.

C   If the initial concentrations of A and B are both doubled, the rate of the reaction will be doubled.

D   As the reaction proceeds, its rate will decrease.

25. Which of the following does **not** apply to the reaction between methane and chlorine?

It involves

A   homolytic fission

B   a chain reaction

C   a carbocation intermediate

D   free radicals.

26. Which of the following amines has the lowest boiling point?

A $C_4H_9NH_2$

B $C_3H_7NHCH_3$

C $C_2H_5NHC_2H_5$

D $C_2H_5N(CH_3)_2$

27. Chlorobenzene, nitrobenzene and ethylbenzene can all be formed from benzene by

A   electrophilic substitution

B   electrophilic addition

C   nucleophilic substitution

D   nucleophilic addition.

28.

X   Y   Z

Which of the following shows the above compounds in order of **increasing** acid strength?

A   X   Y   Z

B   Y   Z   X

C   Z   X   Y

D   X   Z   Y

29. Which of the following has nucleophilic properties?

A   Na

B Br^+

C CH_3^+

D NH_3

30. In the homologous series of amines, increase in chain length from CH_3NH_2 to $C_4H_9NH_2$ is accompanied by

	Volatility	Solubility in water
A	increased	increased
B	decreased	decreased
C	increased	decreased
D	decreased	increased

31. The end-on overlap of two atomic orbitals lying along the axis of a bond is known as

A   hybridisation

B   a sigma bond

C   a pi bond

D   a double bond.

32. Alkenes react with ozone (O_3) to form ozonides which can be hydrolysed to give carbonyl compounds.

$$\begin{array}{c} \diagdown \\ \diagup \end{array} C = C \begin{array}{c} \diagup \\ \diagdown \end{array} \xrightarrow{O_3} \text{OZONIDE} \xrightarrow{\text{hydrolysis}} \begin{array}{c} \diagdown \\ \diagup \end{array} C = O + O = C \begin{array}{c} \diagup \\ \diagdown \end{array}$$

Which of the following alkenes will produce a mixture of propanone and ethanal when acted upon in this way?

A $CH_3CH = CHCH_2CH_3$

B $CH_3CH = CHCH_3$

C $(CH_3)_2C = CH_2$

D $CH_3CH = C(CH_3)_2$

33. Which of the following is the formula for a tertiary halogenoalkane?

A $CHBr_3$

B $(CH_3)_3CBr$

C $(CH_2Br)_3CH$

D $BrCH_2C(CH_3)_3$

34. A compound, X, has the formula C_6H_{12}. X **must** be

A   a hydrocarbon

B   an alkene

C   a cycloalkane

D   hexene.

35. A substance, X, is **readily** oxidised by acidified potassium dichromate solution to give a product which does **not** react with sodium carbonate, **nor** with Tollens' reagent.

Which of the following could represent the structure of X?

A $CH_3CH_2CH_2OH$

B $CH_3CH(OH)CH_3$

C CH_3CCH_3
         $\overset{\|}{O}$

D $CH_3CH_2C\overset{\diagup O}{\underset{\diagdown OH}{}}$

36. Which of the following compounds would liberate one mole of hydrogen gas if one mole of it reacts with two moles of sodium?

A C_2H_5OH

B $HOCH_2CH_2OH$

C CH_3COOH

D CH_3CHO

37. Which of the following will **not** form a derivative with 2,4-dinitrophenylhydrazine?

A $CH_3CH_2C\overset{\diagup O}{\underset{\diagdown H}{}}$

B $CH_3CH_2C\overset{\diagup O}{\underset{\diagdown OH}{}}$

C $CH_3CH_2\overset{\overset{O}{\|}}{C}CH_3$

D $\langle \bigcirc \rangle - C\overset{\diagup O}{\underset{\diagdown H}{}}$

[Turn over

38. Spectral studies of an organic compound indicated the presence of a di-substituted benzene ring, two methyl groups and a molecular weight of 134.

Which of the following is a possible structure for the compound?

A

CH$_3$

COCH$_3$

B

CH$_3$

CHO

CH$_3$

C

CH$_3$CH$_2$

COCH$_3$

D

CH$_3$

CH$_2$CHO

39. Which of the following compounds has a geometric isomer?

A

       H   Cl
       | |
H — C — C — H
       | |
       Cl   H

B

H   Cl
    C = C
H   H

C

H   Cl
    C = C
H   Cl

D

H   Cl
    C = C
Cl   H

40. A simplified mass spectrum of an organic compound is shown.

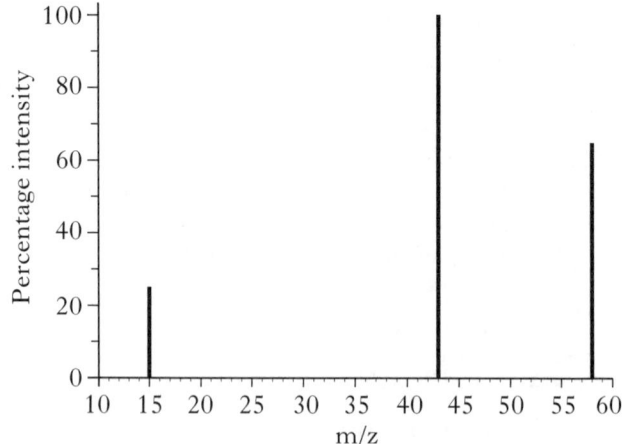

Which of the following compounds produces this spectrum?

A   Propane

B   Propan-1-ol

C   Propan-2-ol

D   Propanone

[END OF SECTION A]

Candidates are reminded that the answer sheet for Section A MUST be placed INSIDE the front cover of your answer book.

SECTION B

60 marks are available in this section of the paper.

All answers must be written clearly and legibly in ink.

Marks

1. Barium carbonate decomposes on heating.

$$BaCO_3(s) \longrightarrow BaO(s) + CO_2(g) \qquad \Delta H^\circ = +266\,kJ\,mol^{-1}$$

   (a) Using the data from the table below, calculate the standard entropy change, ΔS°, in $J\,K^{-1}\,mol^{-1}$, for the reaction.

Substance	Standard entropy, $S^\circ/J\,K^{-1}\,mol^{-1}$
$BaCO_3(s)$	112·0
$BaO(s)$	72·1
$CO_2(g)$	213·8

1

   (b) Calculate the temperature at which the decomposition of barium carbonate just becomes feasible.

3

(4)

2. Consider the following reaction.

$$CS_2(g) + 4H_2(g) \rightleftharpoons CH_4(g) + 2H_2S(g)$$

   At 900 °C the equilibrium concentrations are:

$$[CS_2] = 0·012\,mol\,l^{-1} \qquad [H_2] = 0·0020\,mol\,l^{-1}$$
$$[H_2S] = 0·00010\,mol\,l^{-1} \qquad [CH_4] = 0·0054\,mol\,l^{-1}$$

   (a) Write down the expression for the equilibrium constant, K, for this reaction.

1

   (b) Calculate the value of the equilibrium constant, K, at 900 °C.

1

(2)

3. Fizzy drinks contain carbon dioxide dissolved in water which dissociates, as shown, to produce carbonic acid.

$$H_2O(\ell) + CO_2(aq) \rightleftharpoons H^+(aq) + HCO_3^-(aq) \qquad pK_a = 6·4$$

   (a) What is the Bronsted-Lowry definition of an acid?

1

   (b) Write the formula for the conjugate base in this reaction.

1

   (c) Calculate the pH of a $0·1\,mol\,l^{-1}$ solution of carbonic acid.

2

(4)

[Turn over

Marks

4. The following table of results was obtained for the reaction below.

$$H_2O_2(aq) \quad + \quad 2HI(aq) \quad \longrightarrow \quad 2H_2O(\ell) \quad + \quad I_2(aq)$$

Experiment	$[H_2O_2]/mol\,l^{-1}$	$[HI]/mol\,l^{-1}$	Initial rate/$mol\,l^{-1}\,s^{-1}$
1	$3{\cdot}2 \times 10^{-4}$	$4{\cdot}1 \times 10^{-4}$	$4{\cdot}3 \times 10^{-9}$
2	$6{\cdot}4 \times 10^{-4}$	$4{\cdot}1 \times 10^{-4}$	$8{\cdot}6 \times 10^{-9}$
3	$3{\cdot}2 \times 10^{-4}$	$8{\cdot}2 \times 10^{-4}$	$8{\cdot}6 \times 10^{-9}$
4	$6{\cdot}4 \times 10^{-4}$	$8{\cdot}2 \times 10^{-4}$	$1{\cdot}72 \times 10^{-8}$

(a) Determine the order of this reaction with respect to

   (i) H_2O_2

   (ii) HI.   1

(b) Write the rate equation for the reaction.   1

(c) Calculate a value for the rate constant, k, including the appropriate units.   2

                                                                                               (4)

5. A student set up the following electrochemical cell at 25 °C.

(a) Calculate ΔG, the actual free energy change, in $kJ\,mol^{-1}$, for this cell when the operating emf is $0{\cdot}94\,V$.   3

(b) What changes should the student make to this cell for it to be operating under standard conditions?   1

(c) Calculate the emf of a $Zn(s)\,|\,Zn^{2+}(aq)\,\|\,Cu^{2+}(aq)\,|\,Cu(s)$ cell operating under standard conditions.   1

                                                                                               (5)

Marks

6. In a PPA, 3·43 g of hydrated nickel(II) sulphate, $NiSO_4.6H_2O$, was dissolved in water and made up to $100\,cm^3$ in a standard flask.

   $20·0\,cm^3$ of this solution was titrated against a $0·101\ mol\,l^{-1}$ solution of EDTA using murexide as an indicator. The results are shown below.

	Rough titre	1st titre	2nd titre	3rd titre
Initial burette reading/cm^3	0·0	0·0	24·6	0·0
Final burette reading/cm^3	24·8	24·6	48·8	24·3
Volume of EDTA added/cm^3	24·8	24·6	24·2	24·3

(a) Give a reason why murexide is a suitable indicator in this titration reaction.   1

(b) Calculate the percentage of nickel present in the hydrated salt from these experimental results.   3

(c) The theoretical yield for the percentage of nickel in the hydrated salt is 22·3%. Suggest a reason for the difference between the theoretical percentage of nickel and the answer calculated in part (b).   1

                                                                                    (5)

7. The electronic configuration for cobalt, Co, in its ground state, is

$$1s^2\ 2s^2\ 2p^6\ 3s^2\ 3p^6\ 3d^7\ 4s^2.$$

(a) In terms of s, p and d orbitals, write down the electronic configurations of
   (i) Cu   1
   (ii) Co^{2+} 1
   in their ground states.

(b) (i) Write an equation which represents the second ionisation energy for copper.   1
   (ii) Explain why copper has a higher second ionisation energy than cobalt.   1

                                                                                    (4)

[Turn over

Marks

8. The following table gives information about aluminium chloride and magnesium chloride.

	Aluminium chloride	Magnesium chloride
Action of heat	sublimes at 178 °C	melts at 714 °C
Relative formula mass in gaseous state	267	95
Action of water	reacts to give white fumes	dissolves

   (a) Using the above information determine

   (i) the main type of bonding in aluminium chloride and the main type of bonding in magnesium chloride   1

   (ii) the chemical formula for aluminium chloride in the gaseous state.   1

   (b) What are the white fumes given off when water is added to aluminium chloride?   1

   (c) Aluminium chloride can be used as a catalyst in the reaction between benzene and 2-chloropropane. Draw a structure for an organic product formed.   1

   (d) Many ionic compounds adopt the same lattice structure as either sodium chloride or caesium chloride.

   Sodium chloride has coordination 6:6 whereas caesium chloride has coordination 8:8.

   (i) Why do these chlorides have different lattice structures?   1

   (ii) Which of the above structures is iron(II) oxide most likely to adopt?

   Explain your answer.   1

                                                                                   (6)

9. A common dietary supplement taken by athletes and slimmers is called chromium picolinate $[Cr(pic)_3]$. The structure of the picolinate ion, pic, is

   (a) What feature of the picolinate ion makes it suitable for use as a ligand?   1

   (b) In the body, it is thought that the chromium in $[Cr(pic)_3]$ is changed into chromium(VI) by the action of hydrogen peroxide.

   (i) What is the oxidation state of chromium in $[Cr(pic)_3]$?   1

   (ii) What is the role of hydrogen peroxide in this reaction?   1

   (c) A simpler complex of chromium is $[Cr(CN)_6]^{4-}$. What is its systematic name?   1

                                                                                   (4)

Marks

10. Isoamyl acetate is found naturally as the flavour and scent of bananas.

    Its shortened structural formula is

    $$CH_3COOCH_2CH_2CH(CH_3)_2$$

    (a) (i) To which class of organic compounds does isoamyl acetate belong?   1

         (ii) Apart from flavouring agents, suggest another common use for this class of organic compound.   1

    Isoamyl acetate can be made in the laboratory by reacting ethanoic acid with another substance, **X**.

    (b) (i) Write the systematic name of substance **X**.   1

         (ii) Name the type of reaction which takes place.   1

    (4)

11. Propoxyphene is a pain-killing drug. Its structure is shown below.

    (a) There are two chiral carbons in propoxyphene.
         Referring to the structure above, identify both chiral carbons.   1

    (b) Propoxyphene has a pharmacophore which binds to specific receptors.
         What is meant by the term **pharmacophore**?   1

    (c) Propoxyphene stimulates the body's own natural response to pain.
         What term is used to describe medicines which act in this way?   1

    (3)

    [Turn over

Marks

12. A student designed the following reaction scheme starting from ethyl benzoate.

(a) In a PPA, benzoic acid was prepared from ethyl benzoate.

    (i) Name the type of reaction occurring in step (A).   1

    (ii) What is added to sodium benzoate solution to precipitate out benzoic acid?   1

    (iii) Name the procedure used to purify the benzoic acid.   1

    (iv) Assuming the percentage yield is 70%, what is the minimum mass of ethyl benzoate required to produce at least 4·0 g of benzoic acid?   2

(b) (i) Name the type of reaction occurring in step (B).   1

    (ii) Using structural formulae, outline the mechanism for the reaction occurring in step (C).   2

                                                           (8)

Marks

13. A compound **X** containing only carbon, hydrogen and oxygen, was subjected to elemental analysis. Complete combustion of 1·76 g of **X** gave 3·52 g of carbon dioxide and 1·44 g of water. No other product was formed.

    (a) (i) Calculate the masses of carbon and hydrogen in the original sample and hence deduce the mass of oxygen present.   **2**

    (ii) Show, by calculation, that the empirical formula of compound **X** is C_2H_4O.   **1**

    (b) Given that the relative molecular mass of compound **X** is 88, deduce its molecular formula.   **1**

    (c) A low-resolution proton nmr spectrum of compound **X** is shown.

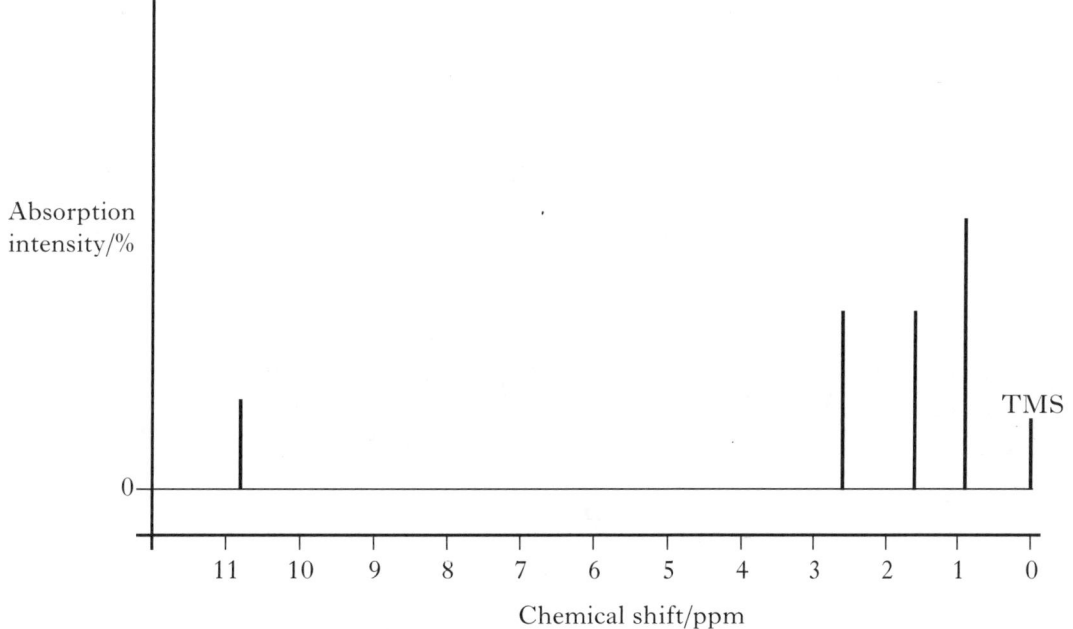

Analysis of this spectrum produced the data shown in the table below.

Chemical shift	Relative area under the peak
0·9	3
1·6	2
2·6	2
10·8	1

    (i) Using information from the table and your answer to (b), draw a structural formula for compound **X** and give its systematic name.   **2**

    (ii) What is the function of "TMS" (tetramethylsilane) in the proton nmr spectrum?   **1**

    (7)

[END OF QUESTION PAPER]

[BLANK PAGE]

[BLANK PAGE]

X012/701

NATIONAL QUALIFICATIONS 2008	FRIDAY, 30 MAY 9.00 AM – 11.30 AM	CHEMISTRY ADVANCED HIGHER

Reference may be made to the Chemistry Higher and Advanced Higher Data Booklet.

SECTION A — 40 marks

Instructions for completion of **SECTION A** are given on page two.

For this section of the examination you must use an **HB pencil**.

SECTION B — 60 marks

All questions should be attempted.

Answers must be written clearly and legibly in ink.

SECTION A

Read carefully

1 Check that the answer sheet provided is for **Chemistry Advanced Higher (Section A)**.

2 For this section of the examination you must use an **HB pencil** and, where necessary, an eraser.

3 Check that the answer sheet you have been given has **your name**, **date of birth**, **SCN** (Scottish Candidate Number) and **Centre Name** printed on it.

   Do not change any of these details.

4 If any of this information is wrong, tell the Invigilator immediately.

5 If this information is correct, **print** your name and seat number in the boxes provided.

6 The answer to each question is **either** A, B, C or D. Decide what your answer is, then, using your pencil, put a horizontal line in the space provided (see sample question below).

7 There is **only one correct** answer to each question.

8 Any rough working should be done on the question paper or the rough working sheet, **not** on your answer sheet.

9 At the end of the exam, put the **answer sheet for Section A inside the front cover of your answer book**.

Sample Question

To show that the ink in a ball-pen consists of a mixture of dyes, the method of separation would be

   A   chromatography

   B   fractional distillation

   C   fractional crystallisation

   D   filtration.

The correct answer is **A**—chromatography. The answer **A** has been clearly marked in **pencil** with a horizontal line (see below).

Changing an answer

If you decide to change your answer, carefully erase your first answer and using your pencil, fill in the answer you want. The answer below has been changed to **D**.

1. An atom has the electronic configuration

$$1s^2 \, 2s^2 \, 2p^6 \, 3s^2 \, 3p^1$$

What is the charge of the most likely ion formed from this atom?

A   −1

B   +1

C   +2

D   +3

2. The electronic configurations, **X** and **Y**, for two uncharged atoms of sodium are as follows.

X $1s^2 \, 2s^2 \, 2p^6 \, 3s^1$
Y $1s^2 \, 2s^2 \, 2p^6 \, 4s^1$

Which of the following statements is true?

A   **X** is an excited state.

B   Both **X** and **Y** have vacant 2d orbitals.

C   Energy is absorbed in changing **Y** to **X**.

D   Less energy is required to ionise **Y** compared to **X**.

3. A Lewis base may be regarded as a substance which is capable of donating an unshared pair of electrons to form a covalent bond.

Which of the following could act as a Lewis base?

A Co^{3+}

B PH_3

C BCl_3

D NH_4^+

4. Silicon can be converted into an n-type semiconductor by adding

A   boron

B   carbon

C   arsenic

D   aluminium.

5. Which of the following statements referring to the structures of sodium chloride and caesium chloride is correct?

A   There are eight chloride ions surrounding each sodium ion.

B   There are eight chloride ions surrounding each caesium ion.

C   The chloride ions are arranged tetrahedrally round the sodium ions.

D   The chloride ions are arranged tetrahedrally round the caesium ions.

6. The transition metal salts, MnF_2, FeF_2 and CoF_2, have identical crystal structures because the metal ions have

A   similar radii

B   similar colours

C   the same nuclear charge

D   the same number of d electrons.

7. Which of the following hydrides, when added to water, would give the most acidic solution?

A   Sodium hydride

B   Magnesium hydride

C   Silicon hydride

D   Sulphur hydride

8. Sodium hydride reacts with sodium sulphate as shown.

$$4NaH \, + \, Na_2SO_4 \, \rightarrow \, 4NaOH \, + \, Na_2S$$

This reaction demonstrates sodium hydride acting as

A   a base

B   an acid

C   a reducing agent

D   an oxidising agent.

[Turn over

9. Three elements, **X**, **Y** and **Z**, are in the same period of the Periodic Table.

   The oxide of **X** is amphoteric, the oxide of **Y** is basic and the oxide of **Z** is acidic.

   Which of the following shows the elements arranged in order of increasing atomic number?

   A **Y, X, Z**

   B **Y, Z, X**

   C **Z, X, Y**

   D **X, Y, Z**

10. Which of the following involves oxidation?

    A $MnO_4^- \rightarrow MnO_4^{2-}$

    B $Ag^+ \rightarrow [Ag(NH_3)_2]^+$

    C $[Fe(CN)_6]^{4-} \rightarrow [Fe(CN)_6]^{3-}$

    D $[Ni(H_2O)_6]^{2+} \rightarrow [Ni(CN)_4]^{2-}$

11. The number of unpaired electrons in a gaseous Ni^{2+} ion is

    A 0

    B 2

    C 4

    D 6.

12. $P + Q \rightleftharpoons R + S$

    At 298 K the equilibrium constant for this reaction is $1 \cdot 2 \times 10^{10}$.

    Which of the following is true?

    A The value of $\Delta S°$ must be positive.

    B The value of $\Delta G°$ must be positive.

    C Adding a catalyst will change the equilibrium constant.

    D Increasing the concentration of P will not change the equilibrium constant.

13.

$$CH_3COOH + C_2H_5OH \rightleftharpoons CH_3COOC_2H_5 + H_2O$$

   The above reaction can be said to have reached equilibrium when

   A the equilibrium constant K is equal to 1

   B the reaction between the acid and the alcohol has stopped

   C the concentrations of the products equal those of the reactants

   D the rate of production of ethyl ethanoate equals its rate of hydrolysis.

14. When sulphur dioxide and oxygen react the following equilibrium is established.

    $$2SO_2(g) + O_2(g) \rightleftharpoons 2SO_3(g)$$

    The equilibrium constant for the reaction is 3300 at 630 °C and 21 at 850 °C.

    Which line in the table is correct for the reaction?

	Sign of ΔH	Product yield as temperature increases
A	+	decreases
B	+	increases
C	−	decreases
D	−	increases

15. $500 \, cm^3$ of $0 \cdot 022 \, mol\,l^{-1}$ hydrochloric acid is mixed with $500 \, cm^3$ of $0 \cdot 020 \, mol\,l^{-1}$ sodium hydroxide solution. The pH of the resulting solution will be

    A 2

    B 3

    C 4

    D 5.

16. The Bronsted-Lowry definition of a base is a substance which acts as a

    A proton donor to form a conjugate acid

    B proton donor to form a conjugate base

    C proton acceptor to form a conjugate acid

    D proton acceptor to form a conjugate base.

17. The mean bond enthalpy of the N–H bond is equal to one third of the value of ΔH for which change of the following changes?

   A $NH_3(g) \rightarrow N(g) + 3H(g)$

   B $2NH_3(g) \rightarrow N_2(g) + 3H_2(g)$

   C $NH_3(g) \rightarrow \frac{1}{2}N_2(g) + 1\frac{1}{2}H_2(g)$

   D $2NH_3(g) + 1\frac{1}{2}O_2(g) \rightarrow N_2(g) + 3H_2O(g)$

18. The entropy of a perfect crystal is zero at

   A $0\,K$

   B $25\,K$

   C $273\,K$

   D $298\,K$.

19. Which of the following reactions results in a **decrease** in entropy?

   A $N_2O_4(g) \rightarrow 2NO_2(g)$

   B $N_2(g) + 3H_2(g) \rightarrow 2NH_3(g)$

   C $CaCO_3(s) \rightarrow CaO(s) + CO_2(g)$

   D $C(s) + H_2O(g) \rightarrow CO(g) + H_2(g)$

20. Which of the following is **not** a required condition for measuring standard electrode potentials?

   A   Volume of 1 litre

   B   Temperature of 298 K

   C   Concentration of $1\,mol\,l^{-1}$

   D   Pressure of 1 atmosphere

21.

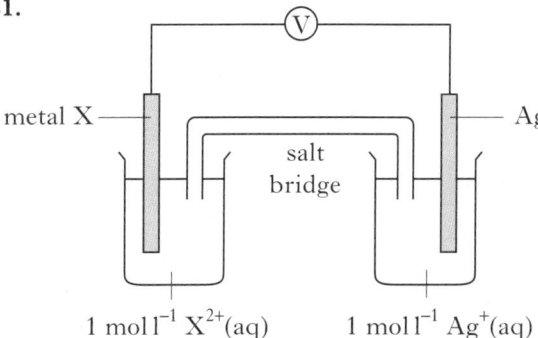

   $1\,mol\,l^{-1}\,X^{2+}(aq)$ $1\,mol\,l^{-1}\,Ag^+(aq)$

   The E° values are

   $$X^{2+}(aq) + 2e^- \rightarrow X(s) \qquad E^\circ = -0.23\,V$$
   $$Ag^+(aq) + e^- \rightarrow Ag(s) \qquad E^\circ = 0.80\,V$$

   In the above cell, which of the following is reduced?

   A $X(s)$

   B $Ag(s)$

   C $X^{2+}(aq)$

   D $Ag^+(aq)$

22. Under standard conditions, the emf of the cell

   $$Al(s)\,|\,Al^{3+}(aq)\,||\,Cu^{2+}(aq)\,|\,Cu(s)$$

   would be

   A $1.34\,V$

   B $2.02\,V$

   C $2.34\,V$

   D $4.38\,V$.

23. For a cell in which the following reaction occurs

   $$X(s) + 2Y^+(aq) \rightarrow X^{2+}(aq) + 2Y(s)$$

   the E° value is $1.5\,V$.

   ΔG° for the reaction, per mole of X, is

   A $-289.5\,kJ\,mol^{-1}$

   B $-144.8\,kJ\,mol^{-1}$

   C $+144.8\,kJ\,mol^{-1}$

   D $+289.5\,kJ\,mol^{-1}$.

[Turn over

24.

The two steps in the reaction mechanism shown can be described as

A   ethene acting as a nucleophile and Br⁻ acting as a nucleophile

B   ethene acting as a nucleophile and Br⁻ acting as an electrophile

C   ethene acting as an electrophile and Br⁻ acting as a nucleophile

D   ethene acting as an electrophile and Br⁻ acting as an electrophile.

25.   In the homologous series of alkanols, increase in chain length from CH_3OH to $C_{10}H_{21}OH$ is accompanied by

A   increased volatility and increased solubility in water

B   increased volatility and decreased solubility in water

C   decreased volatility and decreased solubility in water

D   decreased volatility and increased solubility in water.

26.   Which of the following is **not** caused by hydrogen bonding?

A   The low density of ice compared to water

B   The solubility of methoxymethane in water

C   The higher boiling point of methanol compared to ethane

D   The higher melting point of hydrogen compared to helium

27.   A compound C_3H_8O does **not** react with sodium and is **not** reduced by lithium aluminium hydride. It is likely to be an

A   acid

B   ether

C   alcohol

D   aldehyde.

28.   Which of the following is least acidic?

A CH_3OH

29. Which statement about ethanol and its isomeric ether is true?

They

A have similar volatilities

B have similar infra-red spectra

C form the same products when burned in excess oxygen

D form the same products when reacted with acidified dichromate.

30. When 2-bromobutane is reacted with potassium cyanide and the compound formed is hydrolysed with dilute acid, the final product is

A butanoic acid

B pentanoic acid

C 2-methylbutanoic acid

D 2-methylpropanoic acid.

31. Which of the following compounds would liberate one mole of hydrogen gas if one mole of it reacts with excess sodium?

A C_2H_5OH

B CH_3CHO

C CH_3COOH

D $HOCH_2CH_2OH$

32. Two isomeric esters, **X** and **Y**, have the molecular formula $C_4H_8O_2$. Ester **X** on hydrolysis with sodium hydroxide solution gives CH_3CH_2COONa, and ester **Y** on similar treatment gives CH_3CH_2OH.

Which line in the table shows the correct names of **X** and **Y**?

	X	Y
A	propyl methanoate	ethyl ethanoate
B	methyl propanoate	ethyl ethanoate
C	methyl propanoate	ethyl methanoate
D	propyl methanoate	methyl propanoate

33. A white crystalline compound, soluble in water, was found to react with both dilute hydrochloric acid and sodium hydroxide solution.

Which of the following might it have been?

A C_6H_5OH

B $C_6H_5NH_2$

C C_6H_5COOH

D H_2NCH_2COOH

34. Which of the following amines has the lowest boiling point?

A $C_4H_9NH_2$

B $C_3H_7NHCH_3$

C $C_2H_5NHC_2H_5$

D $C_2H_5N(CH_3)_2$

35. Spectral studies of an organic compound indicated the presence of a di-substituted benzene ring, two methyl groups and a molecular weight of 134.

Which of the following is a possible structure for the compound?

A

B

C

D

[Turn over

36. Which of the following molecules does **not** exhibit optical isomerism?

A — Br
       Br

B   Br — — Br

C — CH(OH)CH$_3$

D — CHBrCHO

37. Which of the following could **not** exist in isomeric forms?

A   C$_2$F$_4$

B   C$_3$H$_6$

C   C$_3$H$_7$Br

D   C$_2$H$_4$Cl$_2$

38. Elemental analysis of an organic compound showed it contained 70·6% carbon, 23·5% oxygen and 5·9% hydrogen by mass.

The structural formula of the compound could be

A   CH$_2$CHCH$_2$COOH

B — OH
       OH

C — CH$_2$CO$_2$H

D — CH$_2$CHO

39. An organic compound with empirical formula, C$_2$H$_4$O, has major peaks at 1715 cm^{-1} and 3300 cm^{-1} in its infrared spectrum.

The structural formula of the compound could be

A   CH$_3$CHO

B   CH$_3$COOH

C   CH$_3$COOCH$_2$CH$_3$

D   CH$_3$CH$_2$CH$_2$COOH.

40. A drug containing a carboxyl group can bind to an amino group on a receptor site in three different ways.

The drug with the following structure

could bind to the same site

A   only by ionic interaction

B   only as a hydrogen-bond donor

C   only as a hydrogen-bond acceptor

D   both as a hydrogen-bond donor and acceptor.

[END OF SECTION A]

Candidates are reminded that the answer sheet for Section A MUST be placed INSIDE the front cover of your answer book.

[Turn over for SECTION B on *Page ten*

SECTION B

60 marks are available in this section of the paper.

All answers must be written clearly and legibly in ink.

Marks

1. (*a*) What is the arrangement of the **electron pairs** around the iodine atom in an IF_5 molecule?

    1

 (*b*) By considering the electron pairs, explain why the bond angle in BF_3 is greater than the bond angle in NF_3.

    1

    (2)

2. An aqueous solution of the compound $[CoCl_2(NH_3)_4]Cl$ gave the following **transmittance** spectrum.

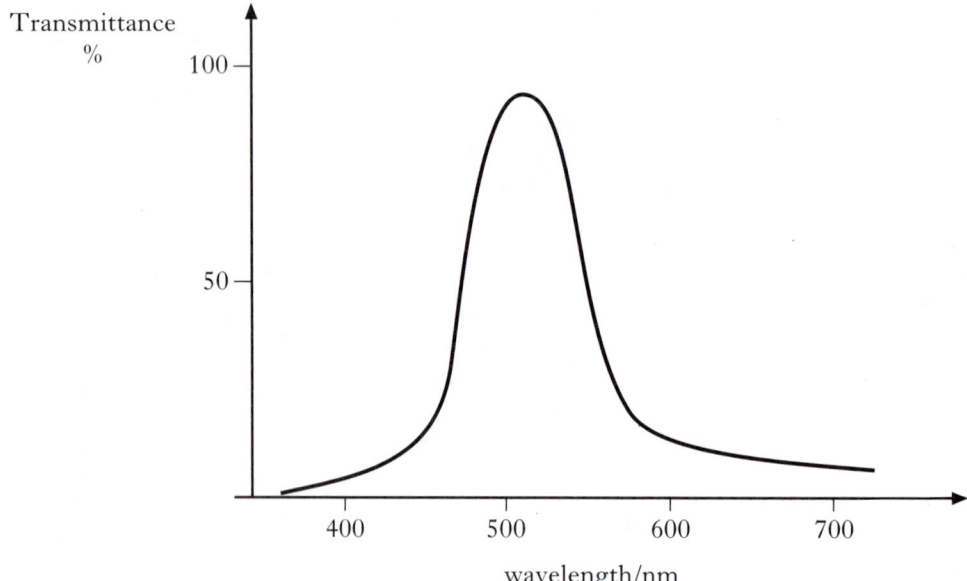

 (*a*) From the above spectrum, deduce the colour of the solution.

    1

 (*b*) The solution contains the complex ion $[CoCl_2(NH_3)_4]^+$.

    (i) What is the oxidation number of cobalt in this complex ion?

    1

    (ii) Name this complex ion.

    1

    (iii) Write down the electronic configuration of cobalt in this complex ion in terms of s, p and d orbitals.

    1

    (4)

3. Some metal salts emit light when heated in a Bunsen flame.

 Lithium nitrate changes the flame colour to crimson.

 Magnesium nitrate has no effect on the flame colour.

 (*a*) Explain, in terms of electrons, why some metal salts emit light when heated in a Bunsen flame.

    1

 (*b*) Suggest why magnesium nitrate has no effect on the flame colour.

    1

 (*c*) Calculate the energy, in kJ mol^{-1}, associated with crimson light of wavelength 671 nm.

    2

    (4)

Marks

4. *cis*-Platin is a highly successful anti-cancer drug. The formula for *cis*-platin is $[Pt(NH_3)_2Cl_2]$.

   (*a*) *cis*-Platin works by forming a complex with parts of a DNA molecule. These parts of the DNA form bonds through nitrogen atoms to $Pt(NH_3)_2$ as shown below.

Parts of DNA molecule

      (i)   Explain why DNA can be classified as a bidentate ligand in this complex.   1

      (ii)   What feature of the DNA makes it suitable as a ligand?   1

  (*b*)   Draw a possible structure for the geometric isomer of *cis*-platin.   1

                                                                   (3)

5. The equation for the decomposition of ammonium dichromate is

$$(NH_4)_2Cr_2O_7(s) \longrightarrow N_2(g) + 4H_2O(\ell) + Cr_2O_3(s)$$

Consider the following data for the reaction at 298 K.

Substance	$\Delta H_f^\circ / kJ\,mol^{-1}$	$S^\circ / J\,K^{-1}\,mol^{-1}$
$(NH_4)_2Cr_2O_7(s)$	−1806	336
$N_2(g)$	0	192
$H_2O(\ell)$	−286	70
$Cr_2O_3(s)$	−1140	81

  (*a*)   For the decomposition of $(NH_4)_2Cr_2O_7$, calculate

      (i) ΔH° 1

      (ii) ΔS° 1

     (iii) ΔG°.   2

  (*b*)   Chromium burns in excess oxygen to form chromium(III) oxide. From the information in the table, deduce a value for the enthalpy of combustion of chromium.   1

                                                                   (5)

[Turn over

Marks

6. Sodium hypochlorite, NaClO, is the active ingredient in household bleach. The concentration of the hypochlorite ion, ClO^-, can be determined in two stages.

   In stage 1, an acidified iodide solution is added to a solution of the bleach and iodine is formed.

   $$ClO^-(aq) + 2I^-(aq) + 2H^+(aq) \rightarrow I_2(aq) + Cl^-(aq) + H_2O(\ell)$$

   In stage 2, the iodine formed is titrated with sodium thiosulphate solution.

   $$2S_2O_3^{2-}(aq) + I_2(aq) \rightarrow 2I^-(aq) + S_4O_6^{2-}(aq)$$

   $10.0\,cm^3$ of a household bleach was diluted to $250\,cm^3$ in a standard flask.

   $25.0\,cm^3$ of this solution was added to excess acidified potassium iodide solution.

   The solution was then titrated with $0.10\,mol\,l^{-1}$ sodium thiosulphate using an appropriate indicator.

   The volume of thiosulphate solution required to reach the end point of the titration was $20.5\,cm^3$.

   (a) Calculate the number of moles of **iodine** which reacted in the titration.   1

   (b) Calculate the concentration, in $mol\,l^{-1}$, of the ClO^- in the original household bleach.   2

   (3)

7. Consider the Born-Haber cycle below which represents the formation of caesium fluoride.

   $$Cs(s) \quad + \quad \tfrac{1}{2}F_2(g) \longrightarrow Cs^+F^-(s)$$

   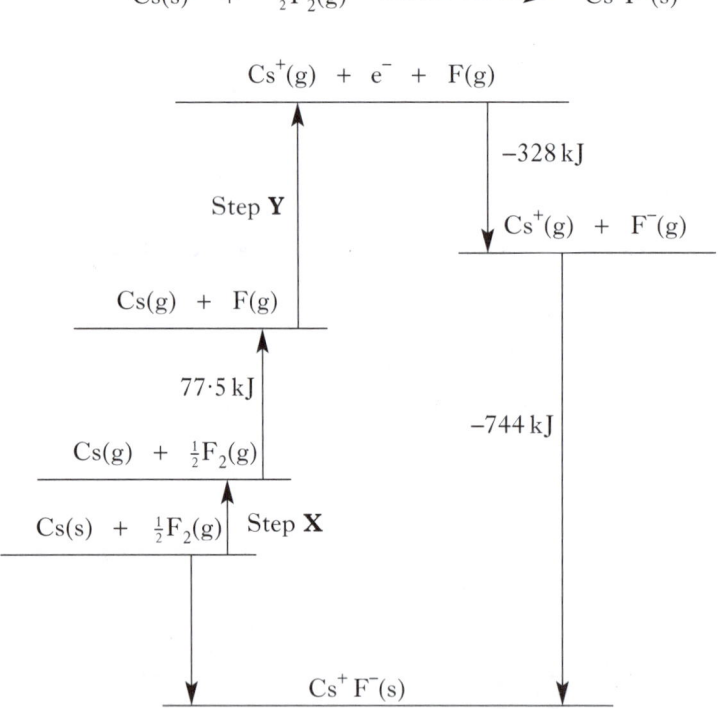

   (a) Use the Data Booklet to find the enthalpy values for Step **X** and Step **Y**.   2

   (b) Name the enthalpy change that has the value $-744\,kJ$ in this cycle.   1

   (c) Use this Born-Haber cycle to calculate the enthalpy of formation of caesium fluoride in $kJ\,mol^{-1}$.   1

   (4)

Marks

8. In a PPA, a student added $50 \, cm^3$ of an aqueous iodine solution to $50 \, cm^3$ of cyclohexane in a separating funnel. After shaking thoroughly, the funnel was left until the following equilibrium was established.

$$I_2(aqueous) \rightleftharpoons I_2(cyclohexane)$$

Two layers were formed, each containing dissolved iodine. $10 \cdot 0 \, cm^3$ of each layer was titrated with sodium thiosulphate solution until the end point was reached.

The cyclohexane layer required $18 \cdot 8 \, cm^3$ of $0 \cdot 025 \, mol \, l^{-1}$ sodium thiosulphate.

The aqueous layer required $10 \cdot 5 \, cm^3$ of $0 \cdot 050 \, mol \, l^{-1}$ sodium thiosulphate.

(a) Which indicator is used to show that the end point has been reached? **1**

(b) Calculate the concentration of iodine in

   (i) the cyclohexane layer **1**

   (ii) the aqueous layer. **1**

(c) Calculate the partition coefficient for iodine between the two solvents. **1**

(d) If $100 \, cm^3$ of cyclohexane had been used instead of the $50 \, cm^3$, what effect would this have on

   (i) the concentration of iodine in the aqueous layer **1**

   (ii) the value of the partition coefficient? **1**

(6)

[Turn over

Marks

9. Hydrofluoric acid, HF, is a weak acid.

$$HF(aq) \ + \ H_2O(\ell) \ \rightleftharpoons \ H_3O^+(aq) \ + \ F^-(aq)$$

A student neutralised $25 \, cm^3$ of hydrofluoric acid solution with sodium hydroxide solution and followed the reaction by measuring the pH.

The graph obtained for this reaction is shown below.

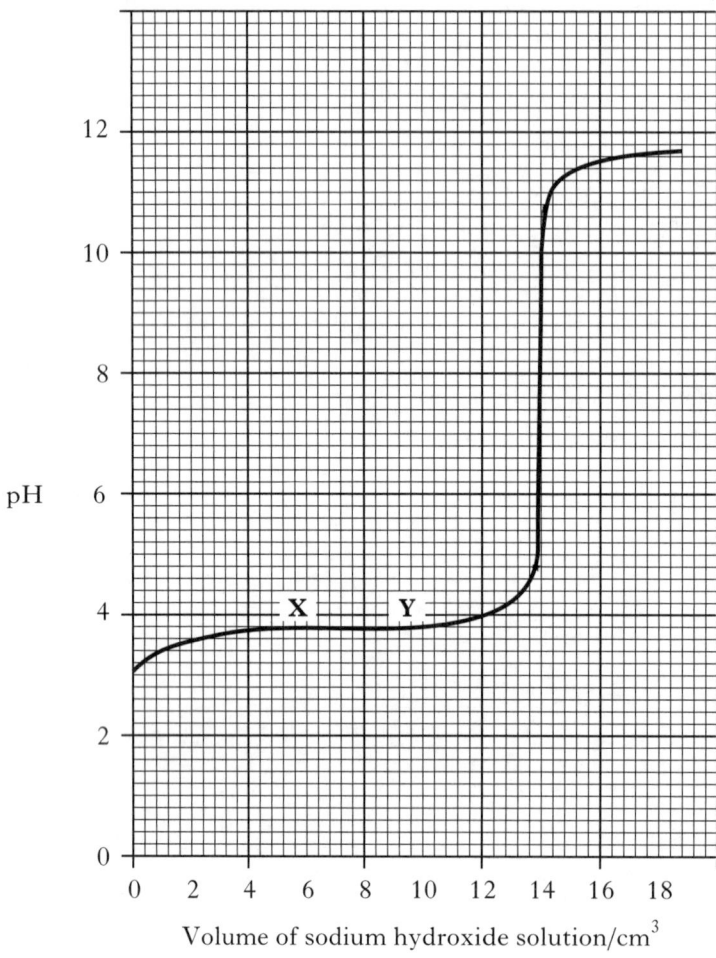

Volume of sodium hydroxide solution/cm^3

(a) Write the expression for the dissociation constant, K_a, of hydrofluoric acid.   1

(b) When exactly half the acid has been neutralised, $pK_a = pH$.

   Using only information from the graph, deduce pK_a and thus calculate K_a for hydrofluoric acid.   2

(c) The region **XY** on the graph is sometimes referred to as the buffer region.

   Apart from HF, what else is present in the solution which enables it to act as a buffer?   1

(d)

Indicator	pK_{In}
Methyl orange	3·7
Alizarin red	6·6
Cresol red	8·0
Alizarin yellow	11·1

   Which of the above indicators could be used to detect the end point of this neutralisation reaction?   1

          (5)

Marks

10. A mixture of butan-1-ol and butan-2-ol can be synthesised from 1-bromobutane in a two stage process.

$$CH_3CH_2CH_2CH_2Br \xrightarrow[\textbf{Stage 1}]{KOH/C_2H_5OH} CH_3CH_2CH=CH_2 \xrightarrow[\textbf{Stage 2}]{H_2O/H^+} \begin{array}{c} \text{butan-1-ol} \\ + \\ \text{butan-2-ol} \end{array}$$

(a) What type of reaction is taking place in **Stage 1**?   1

(b) The bonding in but-1-ene can be described in terms of sp^2 and sp^3 hybridisation and sigma and pi bonds.

   (i) What is meant by sp^2 hybridisation?   1

   (ii) What is the difference in the way atomic orbitals overlap to form sigma and pi bonds?   1

(c) Draw a structural formula for the major product of **Stage 2**.   1

(d) 1-Bromobutane reacts with hydroxide ions in a nucleophilic substitution reaction to produce butan-1-ol. The following results were obtained for this reaction.

Experiment	[1-Bromobutane]/mol l^{-1}	[OH$^-$]/mol l^{-1}	Initial rate/mol l^{-1} s^{-1}
1	0·25	0·10	$3·3 \times 10^{-6}$
2	0·50	0·10	$6·6 \times 10^{-6}$
3	0·50	0·20	$1·3 \times 10^{-5}$

   (i) What is the overall order of this reaction?   1

   (ii) Calculate a value for the rate constant of this reaction, giving the appropriate units.   2

   (iii) Outline the mechanism for this nucleophilic substitution reaction using structural formulae.   2

(9)

11. A student devised the following reaction sequence starting from propan-1-ol, C_3H_7OH.

$$C_3H_7OH \xrightarrow{①} C_3H_7O^- \xrightarrow[②]{C_2H_5Cl} \textbf{Y}$$

$$\downarrow ③$$

$$C_2H_5COOH \xrightarrow[④]{PCl_5} C_2H_5COCl \xrightarrow[⑤]{C_2H_5OH} \textbf{Z}$$

(a) Name a suitable reagent to carry out

   (i) Step ①

   (ii) Step ③.   2

(b) Name **Y**.   1

(c) Draw a structural formula for **Z**.   1

(4)

[Turn over

Marks

12. In a PPA, propanone reacts with 2,4-dinitrophenylhydrazine to make the 2,4-dinitrophenylhydrazone derivative as shown below.

propanone   2,4-dinitrophenylhydrazine   2,4-dinitrophenylhydrazone derivative

(a) What type of reaction is this?   1

(b) The 2,4-dinitrophenylhydrazone derivative formed in the reaction is impure.

    (i) How would the derivative be purified?   1

    (ii) How can the technique of derivative formation be used to identify an unknown ketone?   1

(c) Propanone has an isomer. The shortened structural formula of this isomer is CH_3CH_2CHO.

    (i) Which chemical reagent could be used to distinguish between propanone and this isomer and what would be the result?   1

    (ii) Nuclear magnetic resonance spectroscopy can also be used to distinguish between these two isomers. The proton nmr spectrum for CH_3CH_2CHO is shown.

Relative intensity

TMS

10   9   8   7   6   5   4   3   2   1   0

Chemical shift/ppm

Sketch the proton nmr spectrum you would obtain for propanone.   1

Marks

12. (c) (continued)

(iii) A simplified mass spectrum for propanone is shown below.

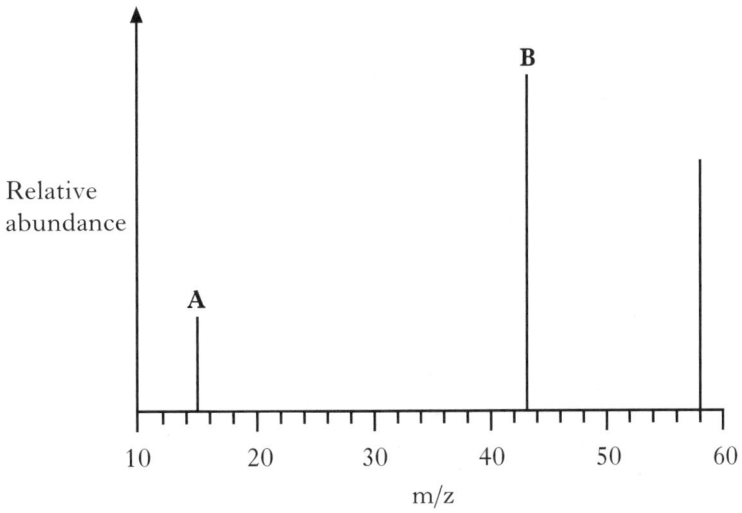

Identify the ion fragments responsible for peaks **A** and **B**.

1

(6)

[Turn over for Question 13 on *Page eighteen*

Marks

13. A student devised the following reaction scheme starting with benzene.

(a) What type of reaction does benzene undergo in reactions ① – ④ ? 1

(b) Name a suitable reagent and catalyst for reaction ① . 1

(c) Reaction ② involves nitration of benzene.
   Which reagents are used to produce the NO_2^+ ion? 1

(d) What is the molecular formula for the product of reaction ③ ? 1

(e) The product of reaction ④ was reacted with bromine in the presence of light.
   Draw a structural formula for an organic product of this reaction. 1

(5)

[END OF QUESTION PAPER]

[BLANK PAGE]

X012/701

NATIONAL QUALIFICATIONS 2009	WEDNESDAY, 3 JUNE 9.00 AM – 11.30 AM	CHEMISTRY ADVANCED HIGHER

Reference may be made to the Chemistry Higher and Advanced Higher Data Booklet .

SECTION A — 40 marks

Instructions for completion of **SECTION A** are given on page two.

For this section of the examination you must use an **HB pencil**.

SECTION B — 60 marks

All questions should be attempted.

Answers must be written clearly and legibly in ink.

SECTION A

Read carefully

1   Check that the answer sheet provided is for **Chemistry Advanced Higher (Section A)**.

2   For this section of the examination you must use an **HB pencil** and, where necessary, an eraser.

3   Check that the answer sheet you have been given has **your name**, **date of birth**, **SCN** (Scottish Candidate Number) and **Centre Name** printed on it.

    Do not change any of these details.

4   If any of this information is wrong, tell the Invigilator immediately.

5   If this information is correct, **print** your name and seat number in the boxes provided.

6   The answer to each question is **either** A, B, C or D. Decide what your answer is, then, using your pencil, put a horizontal line in the space provided (see sample question below).

7   There is **only one correct** answer to each question.

8   Any rough working should be done on the question paper or the rough working sheet, **not** on your answer sheet.

9   At the end of the exam, put the **answer sheet for Section A inside the front cover of your answer book**.

Sample Question

To show that the ink in a ball-pen consists of a mixture of dyes, the method of separation would be

        A   chromatography

        B   fractional distillation

        C   fractional crystallisation

        D   filtration.

The correct answer is **A**—chromatography. The answer **A** has been clearly marked in **pencil** with a horizontal line (see below).

Changing an answer

If you decide to change your answer, carefully erase your first answer and using your pencil, fill in the answer you want. The answer below has been changed to **D**.

1. The diagram shows one of the series of lines in the hydrogen emission spectrum.

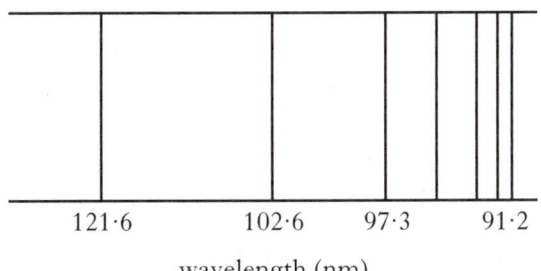

wavelength (nm)

Each line

A represents an energy level within a hydrogen atom

B results from an electron moving to a higher energy level

C lies within the visible part of the electromagnetic spectrum

D results from an excited electron dropping to a lower energy level.

2. Which of the following compounds shows most covalent character?

A CH_4

B NaH

C NH_3

D PH_3

3. In which of the following species is a dative covalent bond present?

A H_3O^+

B H_2O

C OH^-

D O_2

4. Which of the following diagrams best represents the arrangement of electron pairs around the central iodine atom in the I_3^- ion?

A   I —— I —— I

B

C

D

5. When a voltage is applied to an n-type semiconductor, which of the following migrate through the lattice?

A Electrons

B Negative ions

C Positive holes

D Both electrons and positive holes

6. Which of the following compounds would produce fumes of hydrogen chloride when added to water?

A $LiCl$

B $MgCl_2$

C PCl_3

D CCl_4

[Turn over

7. Zinc oxide reacts as shown.

$$ZnO(s) + 2HCl(aq) \rightarrow ZnCl_2(aq) + H_2O(\ell)$$

$$ZnO(s) + 2NaOH(aq) + H_2O(\ell) \rightarrow Na_2Zn(OH)_4(aq)$$

This shows that zinc oxide is

A   basic

B   acidic

C   neutral

D   amphoteric.

8. The correct formula for the tetraamminedichlorocopper(II) complex is

A $[Cu(NH_3)_4Cl_2]^{2-}$

B $[Cu(NH_3)_4Cl_2]$

C $[Cu(NH_3)_4Cl_2]^{2+}$

D $[Cu(NH_3)_4Cl_2]^{4+}$.

9. Which of the following aqueous solutions contains the **greatest** number of **negatively** charged ions?

A $500 \, cm^3 \, 0.10 \, mol \, l^{-1} \, Na_2SO_4(aq)$

B $250 \, cm^3 \, 0.12 \, mol \, l^{-1} \, BaCl_2(aq)$

C $300 \, cm^3 \, 0.15 \, mol \, l^{-1} \, KI(aq)$

D $400 \, cm^3 \, 0.10 \, mol \, l^{-1} \, Zn(NO_3)_2(aq)$

10. When one mole of phosphorus pentachloride was heated to 523 K in a closed vessel, 50% dissociated as shown.

$$PCl_5(g) \rightleftharpoons PCl_3(g) + Cl_2(g)$$

How many moles of gas were present in the equilibrium mixture?

A   0·5

B   1·0

C   1·5

D   2·0

11. Which of the following graphs shows the temperature change as $2 \, mol \, l^{-1}$ sodium hydroxide is added to $25 \, cm^3$ of $2 \, mol \, l^{-1}$ hydrochloric acid?

A

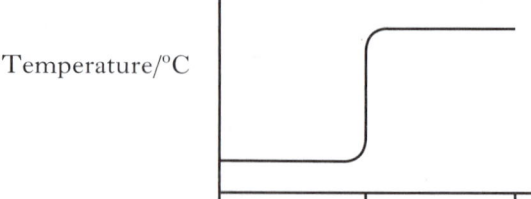

Temperature/°C

Volume of NaOH/cm³

B

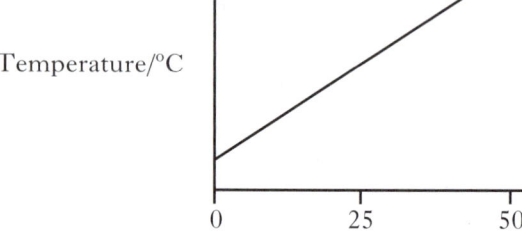

Temperature/°C

Volume of NaOH/cm³

C

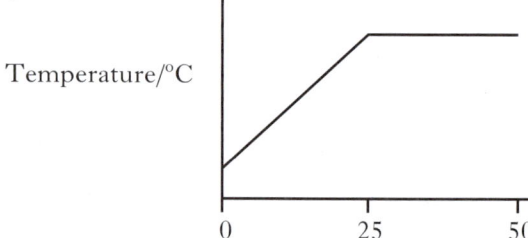

Temperature/°C

Volume of NaOH/cm³

D

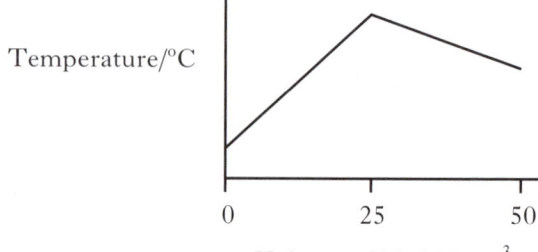

Temperature/°C

Volume of NaOH/cm³

12. In the equilibrium $N_2O_4(g) \rightleftharpoons 2NO_2(g)$ the forward reaction is endothermic.

Which one of the following causes an increase in the value of the equilibrium constant?

A The removal of NO_2

B An increase of pressure

C A decrease of temperature

D An increase of temperature

13. In which of the following separation techniques is partition between two separate phases **not** a part of the process?

A Recrystallisation of benzoic acid from hot water

B Separation of alkanes using gas-liquid chromatography

C Separation of plant dyes using paper chromatography

D Solvent extraction of caffeine from an aqueous solution using dichloromethane

14. An aqueous solution of an organic acid, X, was shaken with chloroform until the following equilibrium was established.

$$X \text{ (water)} \rightleftharpoons X \text{ (chloroform)}$$

X in water

X in chloroform

$25 \cdot 0 \text{ cm}^3$ of the upper layer needed $20 \cdot 0 \text{ cm}^3$ of $0 \cdot 050 \text{ mol l}^{-1}$ NaOH(aq) for neutralisation. $25 \cdot 0 \text{ cm}^3$ of the lower layer needed $13 \cdot 3 \text{ cm}^3$ of $0 \cdot 050 \text{ mol l}^{-1}$ NaOH(aq) for neutralisation.

The value of the partition coefficient is

A $0 \cdot 67$

B $1 \cdot 25$

C $1 \cdot 50$

D $1 \cdot 88$.

15. Which of the following would **not** be suitable to act as a buffer solution?

A Boric acid and sodium borate

B Nitric acid and sodium nitrate

C Benzoic acid and sodium benzoate

D Propanoic acid and sodium propanoate

16. Which of the following $0 \cdot 01 \text{ mol l}^{-1}$ aqueous solutions has the highest pH value?

A Sodium fluoride

B Sodium benzoate

C Sodium propanoate

D Sodium methanoate

17. Which of the following graphs shows the variation in ΔG° with temperature for a reaction which is always feasible?

A

B

C

D

18. When water evaporates from a puddle which of the following applies?

    A ΔH positive and ΔS positive

    B ΔH positive and ΔS negative

    C ΔH negative and ΔS positive

    D ΔH negative and ΔS negative

19. For which of the following reactions would the value of $\Delta G^\circ - \Delta H^\circ$ be approximately zero?

    A $CaCO_3(s) \rightarrow CaO(s) + CO_2(g)$

    B $C(s) + H_2O(g) \rightarrow CO(g) + H_2(g)$

    C $Zn(s) + 2H^+(aq) \rightarrow Zn^{2+}(aq) + H_2(g)$

    D $Cu^{2+}(aq) + Mg(s) \rightarrow Mg^{2+}(aq) + Cu(s)$

20. For the reaction

    $$2NO(g) + Cl_2(g) \rightarrow 2NOCl(g)$$

    the rate equation is

    $$rate = k[NO][Cl_2].$$

    The overall order of this reaction is

    A 1

    B 2

    C 3

    D 5.

21. The following data refer to initial reaction rates obtained for the reaction

    $$X + Y + Z \rightarrow products$$

Run	Relative concentrations			Relative initial rate
	[X]	[Y]	[Z]	
1	1·0	1·0	1·0	0·3
2	1·0	2·0	1·0	0·6
3	2·0	2·0	1·0	1·2
4	2·0	1·0	2·0	0·6

    These data fit the rate equation

    A Rate = $k[X]$

    B Rate = $k[X][Y]$

    C Rate = $k[X][Y]^2$

    D Rate = $k[X][Y][Z]$

22. Which of the following is a propagation step in the chlorination of methane?

    A $Cl_2 \rightarrow Cl\bullet + Cl\bullet$

    B $CH_3\bullet + Cl\bullet \rightarrow CH_3Cl$

    C $CH_3\bullet + Cl_2 \rightarrow CH_3Cl + Cl\bullet$

    D $CH_4 + Cl\bullet \rightarrow CH_3Cl + H\bullet$

23. The hydrolysis of the halogenoalkane $(CH_3)_3CBr$ was found to take place by an S_N1 mechanism.

    The rate-determining step involved the formation of

    A

    B

    C

    D

24. $OH^- + CO_2 \rightarrow HCO_3^-$

$C_2H_4 + Br_2 \rightarrow C_2H_4Br^+ + Br^-$

Which substances act as electrophiles in the above reactions?

A OH^- and Br_2

B OH^- and C_2H_4

C CO_2 and Br_2

D CO_2 and C_2H_4

25.

What is the molecular formula for the above structure?

A $C_{17}H_{11}$

B $C_{17}H_{14}$

C $C_{17}H_{17}$

D $C_{17}H_{20}$

26. Which line in the table is correct for the following hydrocarbon?

$$H_2C = \overset{\overset{\displaystyle CH_3}{|}}{C} - C \equiv CH$$

	Number of σ bonds	Number of π bonds
A	4	3
B	8	5
C	10	2
D	10	3

27. When but-2-ene is shaken with an aqueous solution of chlorine in potassium iodide, the structural formula(e) of the product(s) is/are

A

$$CH_3 - \overset{\overset{\displaystyle I}{|}}{CH} - \overset{\overset{\displaystyle I}{|}}{CH} - CH_3$$

B

$$CH_3 - \overset{\overset{\displaystyle Cl}{|}}{CH} - \overset{\overset{\displaystyle Cl}{|}}{CH} - CH_3$$

C

$$CH_3 - \overset{\overset{\displaystyle Cl}{|}}{CH} - \overset{\overset{\displaystyle I}{|}}{CH} - CH_3 \quad \text{and} \quad CH_2 - \overset{\overset{\displaystyle I}{|}}{CH} - CH_2 - CH_3$$

D

$$CH_3 - \overset{\overset{\displaystyle Cl}{|}}{CH} - \overset{\overset{\displaystyle I}{|}}{CH} - CH_3 \quad \text{and} \quad CH_3 - \overset{\overset{\displaystyle Cl}{|}}{CH} - \overset{\overset{\displaystyle Cl}{|}}{CH} - CH_3$$

28. Which of the following reacts with ethanol to form the ethoxide ion?

A Na(s)

B Na_2O(s)

C NaCl(aq)

D NaOH(aq)

29. Which of the following is **not** a correct statement about ethoxyethane?

A It burns readily in air.

B It is isomeric with butan-2-ol.

C It has a higher boiling point than butan-2-ol.

D It is a very good solvent for many organic compounds.

[Turn over

30. Which of the following esters gives a secondary alcohol when hydrolysed?

A

$$(CH_3)_3C-O-\overset{\overset{\textstyle O}{\|}}{C}-H$$

B

$$CH_3-O-\overset{\overset{\textstyle O}{\|}}{C}-CH(CH_3)_2$$

C

$$(CH_3)_2CH-O-\overset{\overset{\textstyle O}{\|}}{C}-CH_3$$

D

$$(CH_3)_2CHCH_2-O-\overset{\overset{\textstyle O}{\|}}{C}-CH_3$$

31. Which of the following compounds could **not** be oxidised by acidified potassium dichromate solution?

A CH_3CH_2CHO

B CH_3CH_2COOH

C $CH_3CH_2CH_2OH$

D $CH_3CH(OH)CH_3$

32. Which of the following will react with dilute sodium hydroxide solution?

A $CH_3CHOHCH_3$

B $CH_3CH=CH_2$

C CH_3COOCH_3

D $CH_3CH_2OCH_3$

33. Which of the following molecules is planar?

A   Hexane

B   Cyclohexane

C   Chlorobenzene

D   Methylbenzene (toluene)

34. Which of the following compounds is soluble in water and reacts with both dilute hydrochloric acid and sodium hydroxide solution?

A $C_2H_5NH_2$

B $C_6H_5NH_2$

C $C_2H_5NH_3Cl$

D $HOOCCH_2NH_2$

35. Which of the following reactions is least likely to take place?

A

B

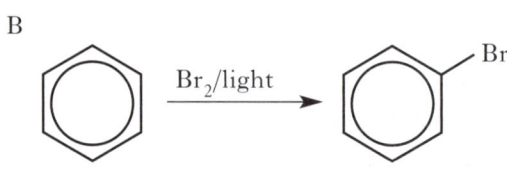

C

D

36. In which of the following pairs does an aqueous solution of the first compound have a higher pH than an aqueous solution of the second?

A   ⬡—OH   and CH_3COOH

B   ⬡—OH   and CH_3CH_2OH

C   ⬡—COOH   and $HOCH_2CH_2OH$

D   ⬡—COOH   and CH_3OH

37. Which of the following bases is the strongest?

A $C_2H_5NH_2$

B $(C_2H_5)_2NH$

C $C_6H_5NH_2$

D $(C_6H_5)_2NH$

38. Which line in the table shows a pair of optical isomers?

A	(CH₃, H, H, CN structure)	(H₃C, H, NC, H structure)
B	(CH₃, Br, H, CN structure)	(H₃C, H, NC, Br structure)
C	(H₃C, H, H, CN structure)	(H, NC, H, CH₃ structure)
D	(H, H₃C, Br, CN structure)	(Br, NC, H, CH₃ structure)

39.

HO—⬡—I

Which atom in the above structure would be located **most** readily using X-ray crystallography?

A   Carbon

B   Hydrogen

C   Iodine

D   Oxygen

40. Antihistamines act by inhibiting the action of the inflammatory agent histamine in the body.

Antihistamines can be described as

A   agonists

B   receptors

C   antagonists

D   pharmacophores.

[END OF SECTION A]

Candidates are reminded that the answer sheet for Section A MUST be placed INSIDE the front cover of your answer book.

SECTION B

60 marks are available in this section of the paper.

All answers must be written clearly and legibly in ink.

Marks

1. A detector in a Geiger counter contains argon which ionises when nuclear radiation passes through it.

    (a) Write the electronic configuration for argon in terms of s and p orbitals.   **1**

    (b) The first ionisation energy of argon is $1530 \, \text{kJ} \, \text{mol}^{-1}$.

        (i) Calculate the wavelength of the radiation, in nm, corresponding to this energy.   **3**

        (ii) Write the equation for the first ionisation of argon.   **1**

    (5)

2. Iron(III) oxide can be reduced to iron using hydrogen.

$$Fe_2O_3(s) + 3H_2(g) \rightarrow 2Fe(s) + 3H_2O(g)$$

Substance	$\Delta H_f^{\circ}/\text{kJ} \, \text{mol}^{-1}$	$S^{\circ}/\text{J} \, \text{K}^{-1} \, \text{mol}^{-1}$
$Fe_2O_3(s)$	−822	90
$H_2(g)$	0	131
$Fe(s)$	0	27
$H_2O(g)$	−242	189

For the reduction of iron(III) oxide with hydrogen, use the data in the table to calculate

(a) the standard entropy change, ΔS° **1**

(b) the standard enthalpy change, ΔH° **1**

(c) the theoretical temperature above which the reaction becomes feasible.   **2**

    (4)

Marks

3. The diagram, which is not drawn to scale, represents the processes involved in a thermochemical cycle.

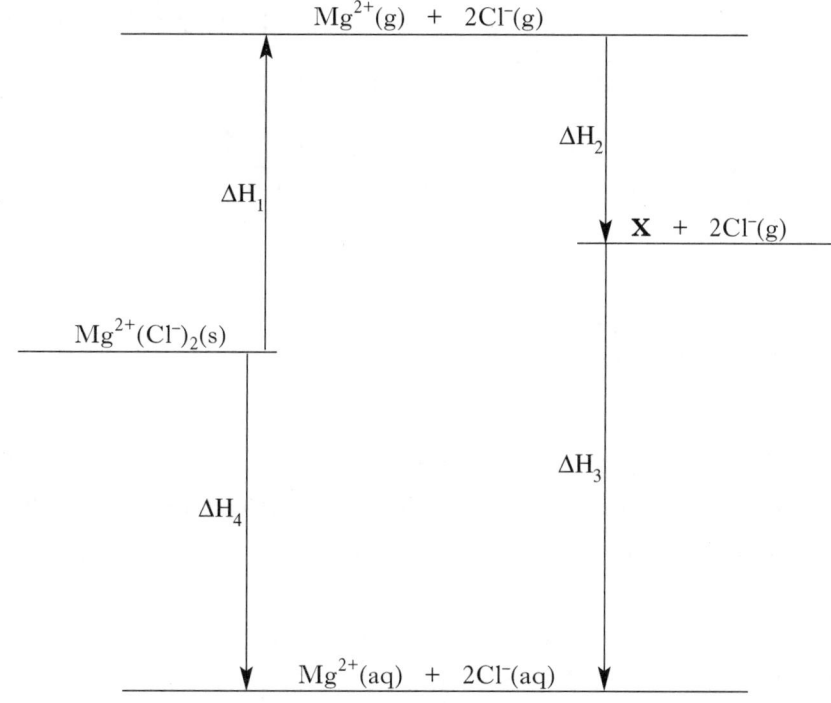

(a) What should be written in place of **X** to complete the diagram? 1

(b) What name is given to the enthalpy change represented by ΔH_1? 1

(c) Calculate ΔH_3 using information from the Data Booklet. 1

(d) Calculate ΔH_4 using information from the Data Booklet. 1

(4)

4. (a) Using the mean bond enthalpy values given in the Data Booklet, calculate the enthalpy change, in kJ mol^{-1}, for the reaction

$$H_2(g) + \tfrac{1}{2}O_2(g) \rightarrow H_2O(g)$$

3

(b) The value given in the Data Booklet for the standard enthalpy of combustion of hydrogen is different to that calculated in part (a).

Give the main reason for this difference. 1

(4)

[Turn over

Marks

5.

Pt electrode

acidified iodate/iodine solution

hydrogen gas

hydrogen reference electrode

The above cell was set up under standard conditions.

(a) What are the three standard conditions required for the hydrogen reference electrode?   **1**

(b) Write an ion-electron equation for the reduction of iodate ions (IO_3^-) to iodine (I_2) in acidic conditions.   **1**

(c) If the E° value for the reduction of IO_3^- to I_2 is 1·19 V, calculate the free energy change $\Delta G°$, in kJ per mole of IO_3^-, for the cell reaction.   **3**

(5)

6. When an ant bites, it injects methanoic acid (HCOOH).

(a) Methanoic acid is a weak acid.

$$HCOOH(aq) \; + \; H_2O(\ell) \; \rightleftharpoons \; HCOO^-(aq) \; + \; H_3O^+(aq)$$

   (i) What is the conjugate base of methanoic acid?   **1**

   (ii) Write the expression for the dissociation constant, K_a, of methanoic acid.   **1**

(b) (i) In a typical bite, an ant injects $3·6 \times 10^{-3}$ g of methanoic acid.

   Assuming that the methanoic acid dissolves in $1·0 \, cm^3$ of water in the body, calculate the concentration of the methanoic acid solution in $mol \, l^{-1}$.   **2**

   (ii) Calculate the pH of this methanoic acid solution.   **2**

(6)

Marks

7. Iodine reacts with propanone as follows.

$$I_2 + CH_3COCH_3 \longrightarrow CH_3COCH_2I + HI$$

A possible mechanism for this reaction is

(a) Write a rate equation for this reaction based on the above mechanism.   **1**

(b) What evidence indicates that the reaction is acid catalysed?   **1**

(c) In a PPA the reaction was followed by withdrawing samples at regular intervals and adding them to sodium hydrogencarbonate solution.

The concentration of iodine in these samples was then determined by titrating with a standard solution of sodium thiosulphate.

  (i) Why were the samples added to the sodium hydrogencarbonate solution?   **1**

  (ii) What indicator is used in the titration and what is the colour change at the end-point of the titration?   **1**

                                                             (4)

[Turn over

Marks

8. Nickel can be determined quantitatively in a number of ways.

   (a) The method used in a PPA is volumetric analysis in which a buffered solution of nickel(II) ions is titrated against a standard solution of a complexing agent.

   Which complexing agent is used?   **1**

   (b) Another way of determining nickel is by colorimetric analysis.

   Why would this be a suitable method of determining nickel(II) ions?   **1**

   (c) A third way of determining nickel depends on the fact that nickel(II) ions form a solid complex with butanedione dioxime.

   butanedione dioxime   insoluble complex

   Using this method, a sample of a nickel(II) salt was accurately weighed and dissolved in water. To this solution, excess butanedione dioxime solution was added. The solid complex formed was filtered, washed and then heated in an oven to constant mass.

   (i) Butanedione dioxime can act as a ligand.

   What property of butanedione dioxime allows it to act as a ligand?   **1**

   (ii) What is the coordination number of the nickel(II) ion in the insoluble complex?   **1**

   (iii) Which type of quantitative analysis has been carried out using this method?   **1**

   (iv) During the process of heating to constant mass, the solid complex is cooled in a desiccator.

   Why is a desiccator used?   **1**

   (6)

Marks

9. Compound **W** reacts in two steps to form compound **Y**.

Y reacts with 2,4-dinitrophenylhydrazine solution (Brady's reagent) to form a yellow precipitate **Z**.

Y does not react with Fehling's solution, nor with Tollens' reagent.

(a)   Identify compound **Y**.   1

(b)   What type of reaction is occurring in step ①?   1

(c)   What property of the yellow precipitate **Z** is measured and how is this used to confirm the identity of **Y**?   1

(d)   Dehydration of compound **X** produces three unsaturated isomers of molecular formula C_4H_8. Two of these are **geometric** isomers.
Draw the structures of both **geometric** isomers and name each one.   2

(5)

10. N-Phenylethanamide can be prepared from benzene in three steps.

(a)   What chemicals are required to react with benzene to bring about step ①?   1

(b)   What type of reaction occurs in step ②?   1

(c)   Suggest a reagent which could be used to bring about step ③.   1

(3)

[Turn over

11. Spectra of an organic compound **A** are shown below.

Mass spectrum of compound A

Infra-red spectrum of compound A

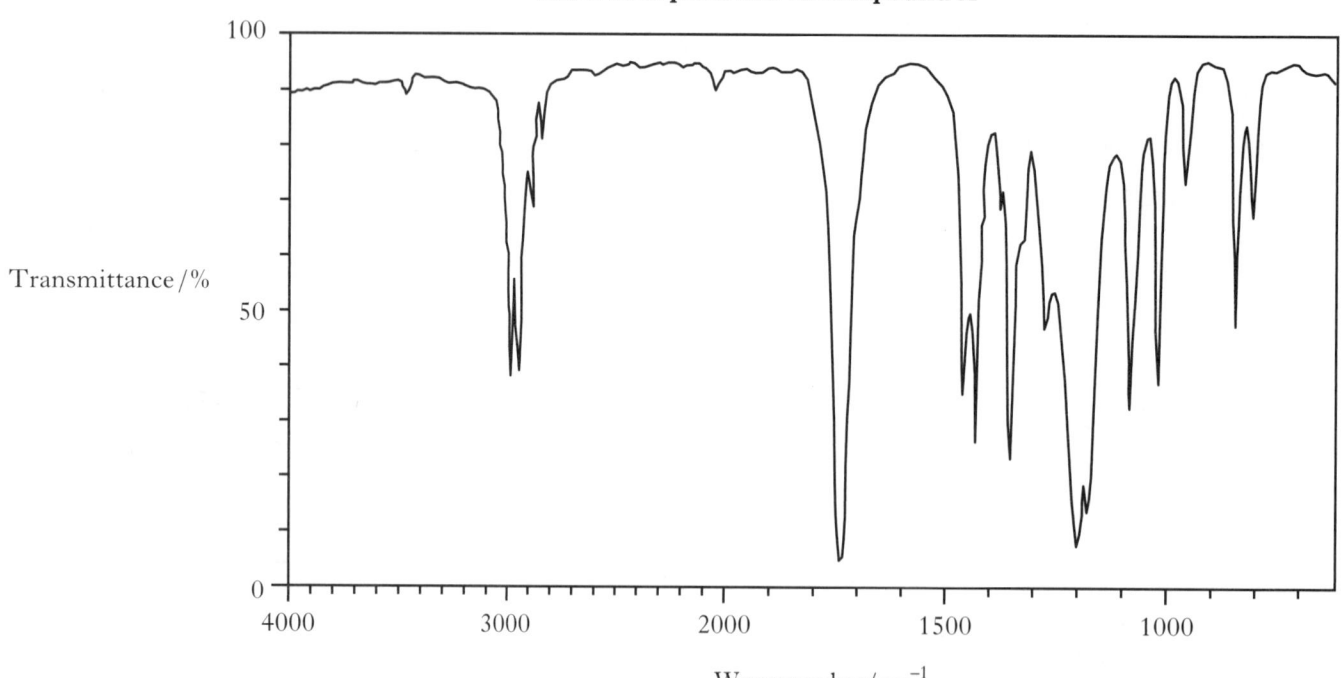

Marks

11. (continued)

Proton nmr spectrum of compound A

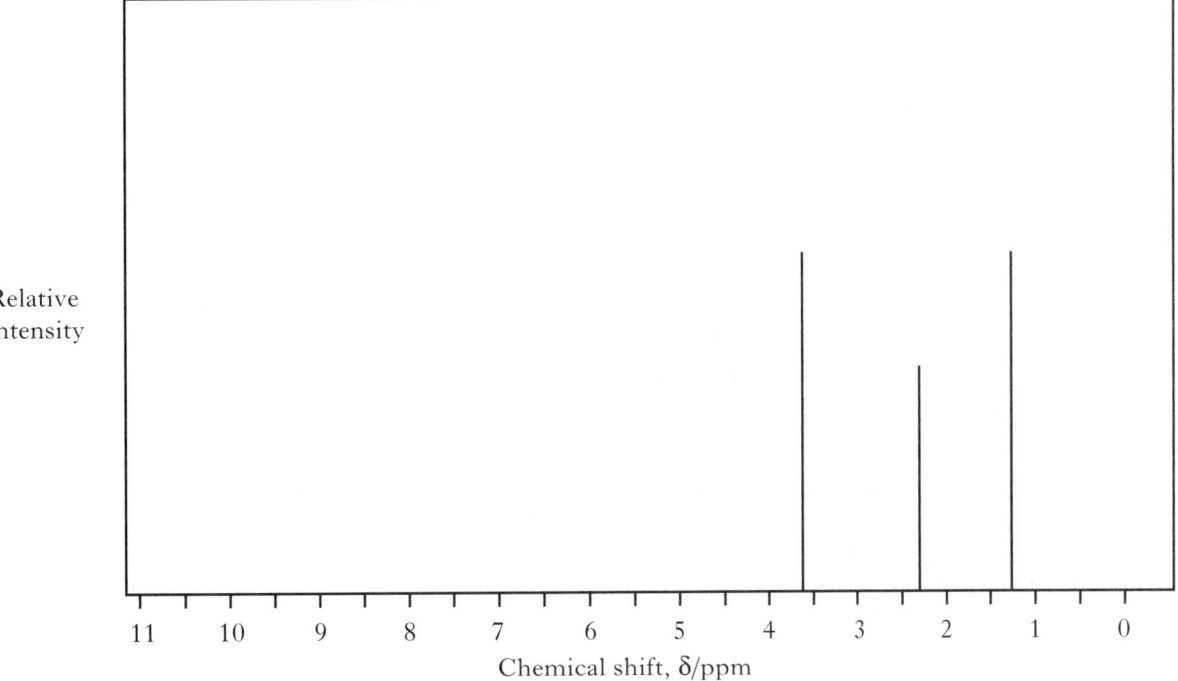

(a) Compound **A** has empirical formula C_2H_4O.

Using this information and the mass spectrum, deduce the molecular formula of **A**.   1

(b) The absorption peak at $1745\,cm^{-1}$ in the infra-red spectrum can be used to help identify **A**.

   (i) Which bond is responsible for this absorption?   1

   (ii) Which type of compound is **A**?   1

(c) Draw the structure of the ion fragment responsible for the peak at m/z 57 in the mass spectrum.   1

(d) Considering all the evidence, including the proton nmr spectrum, name compound **A**.   1

(5)

[Turn over

Marks

12. Many interhalogen compounds exist. Two of these are iodine pentafluoride and iodine heptafluoride.

IF$_5$

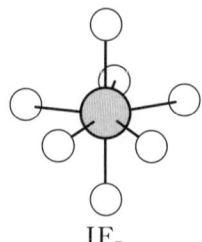

IF$_7$

(a) What are the oxidation states of iodine in iodine pentafluoride and iodine heptafluoride? **1**

(b) Name the shape adopted by the iodine pentafluoride molecule. **1**

(c) In iodine heptafluoride, there are seven I–F bonds in which iodine uses sp^{3}d^3 hybrid orbitals.

Suggest which hybrid orbitals iodine uses in iodine pentafluoride, in which there are five I–F bonds. **1**

(d) Another interhalogen compound, ClF$_5$, exists but ClF$_7$ does not.

Suggest a reason why ClF$_7$ does not exist. **1**

(4)

Marks

13. A superconductor, **X**, with a critical temperature of 95 K, was prepared by heating yttrium oxide, barium carbonate and copper oxide at high temperatures.

    (a) Copy the axes shown and sketch a graph to show how the **electrical resistance** of **X** varies with temperature.

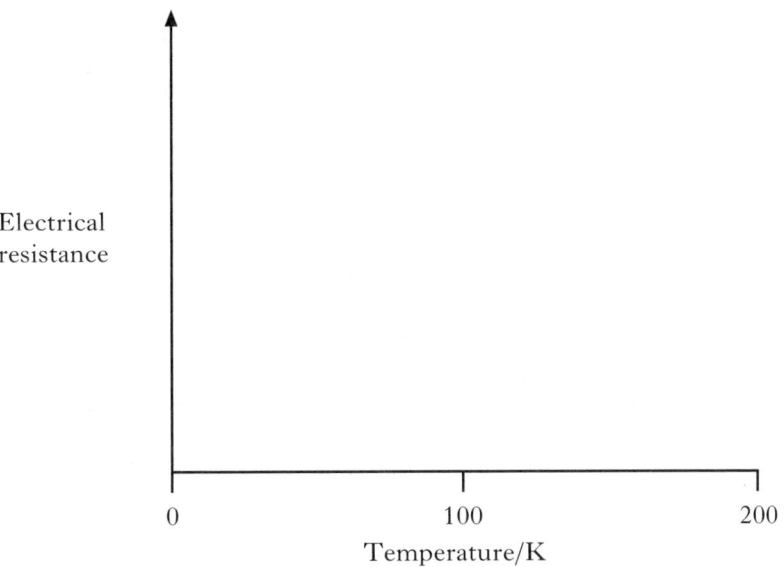

1

    (b) (i) **X** contains 13·4% yttrium, 41·2% barium, 28·6% copper and 16·8% oxygen.

        Assuming that the relative atomic mass of yttrium is 88·9, show by calculation that the empirical formula for **X** is $YBa_2Cu_3O_7$. 2

        (ii) Assuming that the oxidation states of yttrium, barium and oxygen are +3, +2 and −2 respectively, calculate the **average** oxidation state of copper in **X**. 1

        (iii) When all the copper(III) initially present in **X** is reduced to copper(II), compound **Z** is produced. The oxidation states of the other three elements do not change nor does the mole ratio of the **metals**.

        Suggest an empirical formula for **Z**. 1

        (5)

[END OF QUESTION PAPER]

[BLANK PAGE]

ADVANCED HIGHER

2010

[BLANK PAGE]

X012/701

NATIONAL QUALIFICATIONS 2010	WEDNESDAY, 2 JUNE 9.00 AM – 11.30 AM	CHEMISTRY ADVANCED HIGHER

Reference may be made to the Chemistry Higher and Advanced Higher Data Booklet.

SECTION A – 40 marks

Instructions for completion of **SECTION A** are given on page two.

For this section of the examination you must use an **HB pencil**.

SECTION B – 60 marks

All questions should be attempted.

Answers must be written clearly and legibly in ink.

SECTION A

Read carefully

1 Check that the answer sheet provided is for **Chemistry Advanced Higher (Section A)**.

2 For this section of the examination you must use an **HB pencil** and, where necessary, an eraser.

3 Check that the answer sheet you have been given has **your name**, **date of birth**, **SCN** (Scottish Candidate Number) and **Centre Name** printed on it.

    Do not change any of these details.

4 If any of this information is wrong, tell the Invigilator immediately.

5 If this information is correct, **print** your name and seat number in the boxes provided.

6 The answer to each question is **either** A, B, C or D. Decide what your answer is, then, using your pencil, put a horizontal line in the space provided (see sample question below).

7 There is **only one correct** answer to each question.

8 Any rough working should be done on the question paper or the rough working sheet, **not** on your answer sheet.

9 At the end of the examination, put the **answer sheet for Section A inside the front cover of your answer book**.

Sample Question

To show that the ink in a ball-pen consists of a mixture of dyes, the method of separation would be

        A chromatography

        B fractional distillation

        C fractional crystallisation

        D filtration.

The correct answer is **A**—chromatography. The answer **A** has been clearly marked in **pencil** with a horizontal line (see below).

Changing an answer

If you decide to change your answer, carefully erase your first answer and using your pencil, fill in the answer you want. The answer below has been changed to **D**.

A   B   C   D

1. An atom of iron contains 26 electrons.

    Which of the following diagrams below correctly represents the distribution of electrons in the 3d and 4s orbitals in an atom of iron in its ground state?

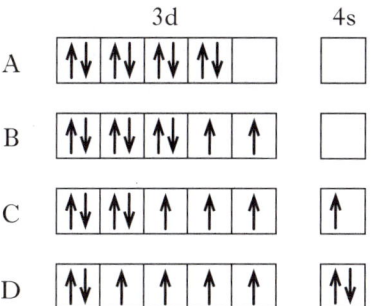

2. An atom has the electronic configuration

    $1s^2 2s^2 2p^6 3s^2 3p^1$

    What is the charge on the most probable ion formed by this element?

    A   +1

    B   +2

    C   +3

    D   +4

3. According to the aufbau principle, electrons fill orbitals in the order

    A   1s 2s 2p 3s 3p 4s 4p 3d

    B   1s 2s 2p 3s 3d 3p 4s 4p

    C   1s 2s 2p 3s 3p 3d 4s 4p

    D   1s 2s 2p 3s 3p 4s 3d 4p.

4. Which line in the graph represents the trend in successive ionisation energies of a Group 2 element?

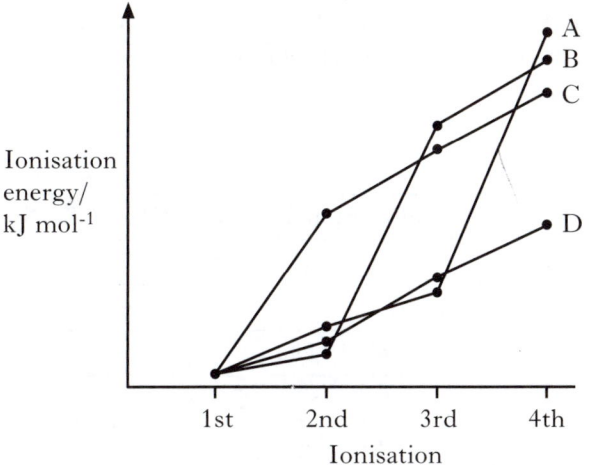

5. Which of the following statements about atomic emission spectroscopy is **incorrect**?

    A   Each element provides a characteristic spectrum.

    B   Visible light is used to promote electrons to higher energy levels.

    C   The lines arise from electron transitions between one energy level and another.

    D   The quantity of the element can be determined from the intensity of radiation transmitted.

6. Which of the following diagrams best represents the arrangement of atoms in the IF_4^- ion?

    Note: a lone pair of electrons is represented by ••

    A

    B

    C

    D

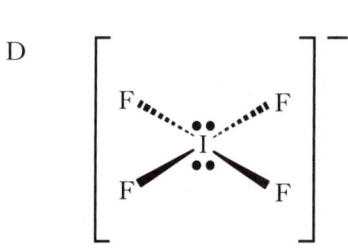

[Turn over

7. Which of the following solids is likely to have the same type of crystal lattice structure as caesium chloride?

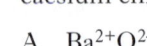

A $Ba^{2+}O^{2-}$

B $Fe^{2+}O^{2-}$

C $Ag^{+}I^{-}$

D $Ni^{2+}O^{2-}$

8. An ionic hydride is added to water.

Which line in the table correctly describes the gas produced and the type of solution formed?

	Gas produced	Type of solution formed
A	hydrogen	acidic
B	hydrogen	alkaline
C	oxygen	acidic
D	oxygen	alkaline

9. An element forms an oxide which is a gas at room temperature.

Which type of bonding is likely to be present in the element?

A Ionic

B Metallic

C Polar covalent

D Non-polar covalent

10. Which of the following oxides would produce the solution with the greatest conductivity when 0·1 mol is added to 250 cm^3 of water?

A SO_2

B CO_2

C Na_2O

D Al_2O_3

11. $2NH_3(g) \rightleftharpoons N_2(g) + 3H_2(g)$ $\Delta H° = 92\,kJ\,mol^{-1}$

The conditions favouring the decomposition of ammonia are

A low pressure and low temperature

B high pressure and low temperature

C low pressure and high temperature

D high pressure and high temperature.

12. An aqueous solution of iodine was shaken with cyclohexane until equilibrium was established.

Some solid iodine was added to the test tube and the contents shaken until equilibrium was re-established.

Which line in the table shows the effects caused by the addition of the solid iodine?

	Concentration of iodine in water	Concentration of iodine in cyclohexane	Partition coefficient
A	increases	increases	no change
B	increases	increases	increases
C	no change	increases	no change
D	increases	no change	increases

13. An acid is a substance which

A donates a proton leaving a conjugate acid

B donates a proton leaving a conjugate base

C accepts a proton leaving a conjugate acid

D accepts a proton leaving a conjugate base.

14. The pH ranges over which some indicators change colour are shown below.

Which line in the table shows the indicator most suitable for the titration of hydrochloric acid with ammonia solution?

	Indicator	pH range
A	Methyl orange	4·2 – 6·3
B	Bromothymol blue	6·0 – 7·6
C	Phenol red	6·8 – 8·4
D	Phenolphthalein	8·3 – 10·0

15. The standard enthalpy of formation of magnesium bromide is the enthalpy change for the reaction

A $Mg^{2+}(g) + 2Br^-(g) \rightarrow Mg^{2+}(Br^-)_2(s)$

B $Mg^{2+}(g) + 2Br^-(g) \rightarrow Mg^{2+}(Br^-)_2(g)$

C $Mg(s) + Br_2(g) \rightarrow Mg^{2+}(Br^-)_2(s)$

D $Mg(s) + Br_2(\ell) \rightarrow Mg^{2+}(Br^-)_2(s)$.

16. The standard enthalpy of combustion of hydrogen is -286 kJ mol^{-1}.

The standard enthalpy of formation of water, in kJ mol^{-1}, is

A -286

B -143

C $+143$

D $+286$.

17. Which of the following enthalpy changes can be measured directly by experiment?

A Bond enthalpy of C–H bond

B Enthalpy of formation of ethane

C Lattice enthalpy of magnesium oxide

D Enthalpy of solution of potassium chloride

18. In which of the following does X represent the bond enthalpy for the O–H bond in water?

A $H_2O(g) \rightarrow O(g) + H_2(g)$ $\Delta H = 2X$

B $H_2O(g) \rightarrow O(g) + 2H(g)$ $\Delta H = 2X$

C $H_2O(g) \rightarrow O(g) + H_2(g)$ $\Delta H = X$

D $H_2O(g) \rightarrow O(g) + 2H(g)$ $\Delta H = X$

19. The enthalpy change for

$Li^+(g) + Br^-(g) \rightarrow Li^+(aq) + Br^-(aq)$

is

A the enthalpy of formation of lithium bromide

B the enthalpy of solution of lithium bromide

C the sum of the hydration enthalpies of lithium and bromide ions

D the sum of the first ionisation energy of lithium and the electron affinity of bromine.

Questions **20** and **21** refer to the Born-Haber cycle below.

20. The enthalpy change which requires the input of most energy is

A ΔH_2

B ΔH_3

C ΔH_4

D ΔH_5.

21. The main enthalpy term which ensures that ΔH_1 is exothermic is

A ΔH_3

B ΔH_4

C ΔH_5

D ΔH_6.

22. Which line in the table shows the correct signs of $\Delta G°$ and $E°$ for a feasible reaction occurring under standard conditions?

	$\Delta G°$	$E°$
A	+	+
B	+	–
C	–	+
D	–	–

23. For the following cell

$Ni(s) \mid Ni^{2+}(aq) \parallel Cu^{2+}(aq) \mid Cu(s)$

the species being reduced is

A $Ni(s)$

B $Ni^{2+}(aq)$

C $Cu^{2+}(aq)$

D $Cu(s)$.

24.

1 mol l^{-1} KI(aq)   1 mol l^{-1} Fe^{3+}(aq)

In the electrochemical cell shown above, operating under standard conditions, the emf produced would be

A   0·23 V

B   0·58 V

C   1·00 V

D   2·88 V.

25. $C_3H_7Cl + C_2H_5O^- \rightarrow C_3H_7OC_2H_5 + Cl^-$

The above reaction is

A   an elimination reaction

B   a nucleophilic addition reaction

C   a nucleophilic substitution reaction

D   an electrophilic substitution reaction.

26.   Which of the following does **not** occur in the reaction between methane and chlorine?

A   A chain reaction

B   Homolytic fission

C   Free radical formation

D   An addition reaction

27.   Which of the following compounds is likely to be the most soluble in water?

A

B

C

D

28.   The sideways overlap of two parallel atomic orbitals lying perpendicular to the axis of the bond is known as

A   hybridisation

B   a pi bond

C   a sigma bond

D   a double bond.

29. If the structure of 3-methylcyclobutene can be represented by

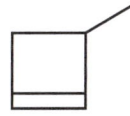

then the structure of 1-ethyl-3-methylcyclopentene will be represented by

A

B

C

D

30. Caryophyllene ($C_{15}H_{24}$) is an unsaturated cyclic hydrocarbon.

Complete hydrogenation of caryophyllene gives a saturated hydrocarbon $C_{15}H_{28}$.

Which line in the table shows the correct numbers of double bonds and rings in caryophyllene?

	Number of double bonds	Number of rings
A	2	1
B	2	2
C	4	2
D	4	4

31. Which of the following is most reactive as a nucleophile?

A Br_2

B CH_3I

C NH_4^+

D NH_3

32. Hydrogen bonding occurs in

A CH_3I

B CH_3OH

C CH_3OCH_3

D CH_3CH_2CHO.

33. Cinnamaldehyde, which can be extracted from cinnamon, has the structure:

Cinnamaldehyde will **not** react with

A sodium metal

B bromine solution

C lithium aluminium hydride

D acidified potassium dichromate.

34. Which of the following compounds would be produced by passing ammonia gas into dilute ethanoic acid?

A CH_3CONH_2

B $CH_3COO^-NH_4^+$

C NH_2CH_2COOH

D $CH_3CH_2NH_3^+Cl^-$

[**Turn over**

35. Secondary amines react with carbonyl compounds to form unsaturated amines known as enamines as shown

an enamine

Which carbonyl compound would react with $(CH_3)_2NH$ to form the enamine with the following structure?

A   Propanal

B   Propanone

C   Butanal

D   Butanone

36.

$$\text{benzene} \xrightarrow{HNO_3 / H_2SO_4} \text{Product } \mathbf{X}$$

Which line in the table is correct for the reaction above?

	Type of reaction	Product **X**
A	electrophilic substitution	(benzene ring)—NO_2
B	electrophilic substitution	(benzene ring)—SO_3H
C	nucleophilic substitution	(benzene ring)—NO_2
D	nucleophilic substitution	(benzene ring)—SO_3H

37. Which of the following statements about the benzene molecule is **not** true?

A   It is planar.

B   It has empirical formula CH.

C   It is readily attacked by bromine.

D   Its C—C bonds are equal in length.

38. Which of the following could **not** exist in isomeric forms?

A C_3H_6

B C_3H_8

C C_3H_7Br

D $C_2H_4Cl_2$

39. Which of the following causes the separation of the ions in a mass spectrometer?

A   A magnetic field

B   A vacuum pump

C   An ionisation chamber

D   Electron bombardment

40. Which of the following compounds is most likely to show an infra-red absorption at 2725 cm^{-1}?

A $$CH_3 - \overset{\overset{\displaystyle O}{\|}}{C} - CH_3$$

B $HOCH_2CH = CH_2$

C $$CH_3 - CH_2 - C\overset{\displaystyle \nearrow O}{\underset{\displaystyle \searrow H}{}}$$

D $CH_3 - O - CH = CH_2$

SECTION B

60 marks are available in this section of the paper.

All answers must be written clearly and legibly in ink.

Marks

1. The first argon compound was prepared by shining light of wavelength 160 nm onto a mixture of argon and hydrogen fluoride at a temperature of 7·5 K. The hydrogen fluoride reacted with the argon to form HArF.

   (a) Calculate the energy, in kJ mol^{-1}, associated with light of wavelength 160 nm.   2

   (b) Supposing HArF is covalent,

      (i) predict the total number of electron pairs, bonding and non-bonding, which surround the Ar atom in the HArF molecule.   1

      (ii) what shape do the electron pairs around the Ar atom in an HArF molecule adopt?   1

                                                                                              (4)

2. Complex ions **A** and **B** are isomeric and have the formula $[Cr(H_2O)_4Cl_2]^+$.

   (a) Calculate the oxidation number of chromium in the complex ion.   1

   (b) Name the complex ion.   1

   (c) The structural formula for complex ion **A** is

   Draw the structural formula for complex ion **B**.   1

                                                                                              (3)

[Turn over

Marks

3. The Thermit process can be used to extract iron from iron(III) oxide.

$$2Al(s) + Fe_2O_3(s) \rightarrow 2Fe(s) + Al_2O_3(s)$$

Substance	Standard enthalpy of formation, $\Delta H°/kJ\ mol^{-1}$	Standard entropy, $S°/J\ K^{-1}\ mol^{-1}$
$Al(s)$	0	28·0
$Fe_2O_3(s)$	-824	87·0
$Fe(s)$	0	27·0
$Al_2O_3(s)$	-1676	51·0

For the Thermit process, use the data in the table to calculate

(a) the standard enthalpy change, $\Delta H°$ **1**

(b) the standard entropy change, $\Delta S°$ **1**

(c) the standard free energy change, $\Delta G°$.   **2**

(4)

Marks

4. In a PPA, a sample of steel was treated in a sequence of reactions to determine the manganese content.

The absorbance of a sample of the permanganate solution formed was analysed using a colorimeter fitted with a 520 nm filter. Optically matched cuvettes were used throughout.

(a) (i) Apart from the steel dissolving in the hot nitric acid, what would have been observed at step one? **1**

(ii) In step two, Mn^{2+}(aq) is converted into MnO_4^-(aq). What is the role of the potassium periodate, KIO_4? **1**

(iii) Why was a 520 nm filter used? **1**

(b) A series of standard permanganate solutions were used to produce the calibration graph below.

Concentration of permanganate solution/mol l^{-1}

The results of the experiment are shown below.

Mass of steel used	=	0·19 g
Absorbance of permanganate solution	=	0·25
Total volume of permanganate solution	=	100 cm³

Use the graph and the results to calculate the percentage, by mass, of manganese in the sample of steel. **3**

(6)

Marks

5. The nitrate ion has three equivalent resonance structures. One of these structures is shown below.

(a) Draw a similar diagram to show one of the other two resonance structures.   1

(b) The formal charge on an atom in a resonance structure can be found using the expression

$$\text{Formal charge} = \left(\begin{array}{c} \text{Group in} \\ \text{Periodic Table} \end{array} \right) - \left(\begin{array}{c} \text{Number of lone pair} \\ \text{electrons} \end{array} \right) - \frac{1}{2} \left(\begin{array}{c} \text{Number of bonding} \\ \text{electrons} \end{array} \right).$$

Use this expression to find the formal charge on atoms (b), (c) and (d) shown in the table below.

Resonance structure	Atom	Formal charge
	(a)	+1
	(b)	?
	(c)	?
	(d)	?

2

(3)

Marks

6. The formula of potassium hydrogen oxalate can be written as $K_xH_y(C_2O_4)_z$.

In an experiment to determine the values of **x, y** and **z**, 4·49 g of this compound was dissolved in water and the solution made up to one litre.

(a) 20·0 cm³ of the solution was pipetted into a conical flask and then titrated with 0·0200 mol l⁻¹ acidified potassium permanganate at 60 °C. The average titre volume was 16·5 cm³.

The equation for the reaction taking place in the conical flask is

$$5C_2O_4^{2-} + 16H^+ + 2MnO_4^- \rightarrow 2Mn^{2+} + 10CO_2 + 8H_2O$$

   (i) What colour change would indicate the end point of the titration?   1

   (ii) From the titration result, calculate the number of moles of oxalate ions, $C_2O_4^{2-}$, in 20·0 cm³ of the solution.   1

   (iii) Calculate the mass of oxalate ions in one litre of the solution.   1

   (iv) Using another analytical procedure, 4·49 g of potassium hydrogen oxalate was found to contain 0·060 g of hydrogen.

   Use this information with the answer to (a)(iii) to calculate the mass of potassium in this sample.   1

(b) Calculate the values of **x, y** and **z**.   2

      (6)

7. The expression for the equilibrium constant of an esterification reaction is

$$K = \frac{[CH_3COOCH_2CH_3]\,[H_2O]}{[CH_3COOH]\,[CH_3CH_2OH]}$$

(a) Write the chemical equation for this esterification reaction.   1

(b) In an experiment to determine the value of the equilibrium constant, 0·70 moles of ethanoic acid and 0·68 moles of ethanol were mixed in a conical flask. The flask was stoppered to prevent the contents escaping and then placed in a water bath at 50 °C.

At equilibrium the mixture contained 0·24 moles of ethanoic acid.

   (i) Why is it important to prevent the contents of the flask escaping?   1

   (ii) Calculate K at 50 °C.   3

      (5)

8. Nicotinic acid is used in the treatment of high cholesterol levels. A structural formula for nicotinic acid is

(a) Write an equation to show the dissociation of nicotinic acid in water.   1

(b) The K_a value of nicotinic acid is $1·4 \times 10^{-5}$.

   Calculate the concentration of a nicotinic acid solution which has a pH of 3·77.   3

      (4)

Marks

9. The rate equation for the reaction between nitrogen dioxide and fluorine is

$$Rate = k[NO_2][F_2]$$

A proposed reaction mechanism is

Step one $\quad NO_2 + F_2 \rightarrow NO_2F + F$

Step two $\quad NO_2 + F \rightarrow NO_2F$.

(a) Which step in the proposed reaction mechanism would be **faster**?   1

(b) Write a balanced equation for the overall reaction.   1

(c) What is the overall order of the reaction?   1

(d)

Experiment	$[NO_2]$/mol l^{-1}	$[F_2]$/mol l^{-1}	Initial rate/ mol l^{-1} s^{-1}
1	0·001	0·003	$1·2 \times 10^{-4}$
2	0·006	0·001	$2·4 \times 10^{-4}$
3	0·002	0·004	$3·2 \times 10^{-4}$

Use the data in the table to calculate a value for the rate constant, k, including the appropriate units.   2

(5)

10. Alkenes can be prepared from alcohols.

In a PPA, 22·56 g of cyclohexanol was dehydrated using an excess of concentrated phosphoric acid. The reaction mixture was then distilled. The crude cyclohexene was added to a separating funnel containing a solution which was used to wash the cyclohexene and improve the separation of the aqueous and organic layers. The organic layer was separated and treated with anhydrous calcium chloride before it was distilled to yield 6·52 g of pure cyclohexene.

(a) Why was concentrated phosphoric acid used as the dehydrating agent rather than concentrated sulphuric acid?   1

(b) Name the solution that the crude cyclohexene was added to in the separating funnel.   1

(c) What was the function of the anhydrous calcium chloride?   1

(d) The relative formula masses of cyclohexanol and cyclohexene are 100 and 82 respectively.

Calculate the percentage yield of cyclohexene.   2

(5)

Marks

11. Consider the following reaction scheme.

(a) Explain why but-2-ene exhibits geometric isomerism yet its structural isomer but-1-ene does not.   1

(b) But-2-ene undergoes electrophilic addition to form (B).

Draw a structure for the carbocation intermediate formed in this electrophilic addition reaction.   1

(c) Name a reagent used to convert (B) to (C).   1

(d) Name a catalyst required in converting (D) to (E).   1

(e) Draw a structural formula for ester (G).   1

(5)

[Turn over

Marks

12. Consider the following reaction sequence.

(a) Name compound **A**. 1

(b) To which class of organic compounds does compound **B** belong? 1

(c) Name the type of reaction taking place in converting compound **B** into lactic acid. 1

(d) Lactic acid in the form of lactate ions is dehydrogenated in the liver by the enzyme, lactate dehydrogenase.

The diagram shows how one of the optical isomers of the lactate ion binds to an active site of lactate dehydrogenase.

(i) Which type of intermolecular force is involved when the methyl group of the lactate ion binds to the hydrophobic region of the active site? 1

(ii) Draw a structure for the other optical isomer of the lactate ion. 1

(iii) Explain why this other optical isomer of the lactate ion cannot bind as efficiently to the active site of lactate dehydrogenase. 1

(6)

Marks

13. Compound **A** has molecular formula $C_4H_{10}O$.

(*a*) To which two classes of organic compounds could **A** belong?

2

(*b*) Compound **A** reacts with acidified potassium dichromate solution to form **B** which has molecular formula C_4H_8O.

The proton nmr spectrum of **B** shows three peaks. Analysis of this spectrum produces the following data.

Peak	Chemical shift/ppm	Relative area under peak
1	0·95	3
2	2·05	3
3	2·35	2

Considering all the evidence above:

(i) draw a structural formula for **B**;

1

(ii) name **A**.

1

(4)

[END OF QUESTION PAPER]

[BLANK PAGE]

[BLANK PAGE]

X012/701

NATIONAL
QUALIFICATIONS
2011

THURSDAY, 26 MAY
9.00 AM – 11.30 AM

CHEMISTRY
ADVANCED HIGHER

Reference may be made to the Chemistry Higher and Advanced Higher Data Booklet .

SECTION A – 40 marks

Instructions for completion of **SECTION A** are given on page two.

For this section of the examination you must use an **HB pencil**.

SECTION B – 60 marks

All questions should be attempted.

Answers must be written clearly and legibly in ink.

SECTION A

Read carefully

1 Check that the answer sheet provided is for **Chemistry Advanced Higher (Section A)**.

2 For this section of the examination you must use an **HB pencil** and, where necessary, an eraser.

3 Check that the answer sheet you have been given has **your name**, **date of birth**, **SCN** (Scottish Candidate Number) and **Centre Name** printed on it.

   Do not change any of these details.

4 If any of this information is wrong, tell the Invigilator immediately.

5 If this information is correct, **print** your name and seat number in the boxes provided.

6 The answer to each question is **either** A, B, C or D. Decide what your answer is, then, using your pencil, put a horizontal line in the space provided (see sample question below).

7 There is **only one correct** answer to each question.

8 Any rough working should be done on the question paper or the rough working sheet, **not** on your answer sheet.

9 At the end of the exam, put the **answer sheet for Section A inside the front cover of your answer book**.

Sample Question

To show that the ink in a ball-pen consists of a mixture of dyes, the method of separation would be

   A chromatography

   B fractional distillation

   C fractional crystallisation

   D filtration.

The correct answer is **A**—chromatography. The answer **A** has been clearly marked in **pencil** with a horizontal line (see below).

Changing an answer

If you decide to change your answer, carefully erase your first answer and using your pencil, fill in the answer you want. The answer below has been changed to **D**.

1. Which of the following lines on the graph represents the trend in successive ionisation energies of a Group 3 element?

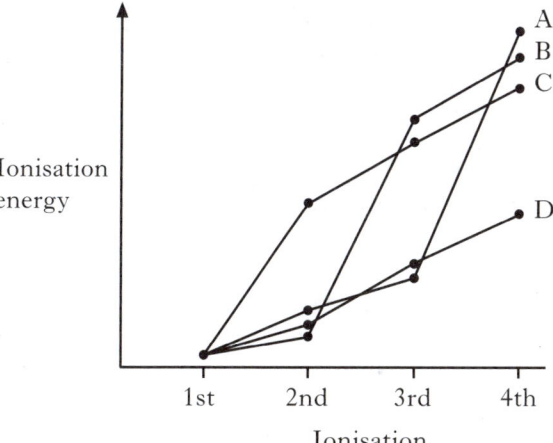

2. In colorimetry, as the concentration of a coloured solution decreases

   A   the absorbance increases

   B   the absorbance decreases

   C   the radiation wavelength increases

   D   the radiation wavelength decreases.

3. Which of the following molecules has the greatest number of non-bonding electron pairs (lone pairs)?

   A   H—C—Cl (with H above and H below C)

   B   H—C—O—H (with H above and H below C)

   C   H—C—N (with H above and H below C, and H above and H below N)

   D   H—C=O (with H above C)

4. What is the change in the three-dimensional arrangement of the bonds round the P atom in the following reaction?

   $$PF_5 \rightarrow PF_3 + F_2$$

   A   Tetrahedral to pyramidal

   B   Octahedral to trigonal planar

   C   Trigonal bipyramidal to pyramidal

   D   Trigonal bipyramidal to trigonal planar

5. The ratio of the ionic radii in sodium chloride is approximately 1:2, whereas in caesium chloride it is approximately 1:1. A compound XY contains X^+ ions with a radius of 133 pm and Y^- ions with a radius of 220 pm.

   In a crystal of XY, how many Y^- ions surround each X^+ ion as its nearest neighbour?

   A   1

   B   2

   C   6

   D   8

6. An example of a p-type semiconductor is silicon doped with

   A   carbon

   B   arsenic

   C   aluminium

   D   phosphorus.

7. Which of the following solid oxides would **not** lower the pH when added to sodium hydroxide solution?

   A Li_2O

   B SiO_2

   C P_4O_{10}

   D Al_2O_3

[Turn over

8. Which of the following is least likely to produce fumes of hydrogen chloride when added to water?

   A PCl_5

   B $SiCl_4$

   C $AlCl_3$

   D $MgCl_2$

9. A white solid gives an orange-yellow flame colour. When added to water, hydrogen gas is released and an alkaline solution is formed.

   The solid could be

   A   sodium oxide

   B   calcium oxide

   C   sodium hydride

   D   calcium hydride.

10. Which of the following ions is **least** likely to be coloured?

   A $Ti(H_2O)_6^{3+}$

   B $Cr(NH_3)_6^{3+}$

   C $Ni(H_2O)_6^{2+}$

   D $Zn(NH_3)_4^{2+}$

11. What volume of $0.25\,mol\,l^{-1}$ calcium nitrate is required to make, by dilution with water, $500\,cm^3$ of a solution with a **nitrate** ion concentration of $0.1\,mol\,l^{-1}$?

   A $50\,cm^3$

   B $100\,cm^3$

   C $200\,cm^3$

   D $400\,cm^3$

12. Hydrogen for use in ammonia production is produced by the endothermic reaction:

   $$CH_4(g) + H_2O(g) \rightleftharpoons CO(g) + 3H_2(g)$$

   Which of the following will increase the equilibrium yield of hydrogen?

   A   Decrease the methane concentration

   B   Decrease the temperature

   C   Decrease the pressure

   D   Add a catalyst

13. The reaction

   $$CO(g) + 3H_2(g) \rightleftharpoons CH_4(g) + H_2O(g)$$

   has an equilibrium constant of 3·9 at 950 °C.

   The equilibrium concentrations of $CO(g)$, $H_2(g)$ and $H_2O(g)$ are given in the table.

Substance	Equilibrium concentration/mol l^{-1}
$CO(g)$	0·500
$H_2(g)$	0·100
$H_2O(g)$	0·040

   What is the equilibrium concentration of $CH_4(g)$, in mol l^{-1}, at 950 °C?

   A   0·049

   B   0·200

   C   4·90

   D   20·0

14.

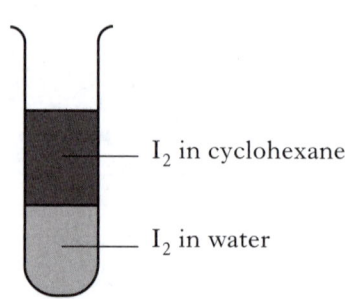

   The partition coefficient for the above system can be altered by

   A   adding more iodine

   B   adding more cyclohexane

   C   changing the temperature

   D   shaking the mixture thoroughly.

15. Gas liquid chromatography could be used to separate a mixture of hydrocarbons. The mixture is passed through a column packed with silica particles coated in a non-polar liquid. Helium can be used to carry the mixture through the column.

Which line in the table identifies correctly the stationary and mobile phases in this chromatographic separation?

	Stationary phase	Mobile phase
A	silica	helium
B	silica	non-polar liquid
C	non-polar liquid	helium
D	non-polar liquid	hydrocarbon mixture

16. Under certain conditions liquid ammonia ionises as shown:

$$2NH_3 \rightleftharpoons NH_4^+ + NH_2^-$$

Which line in the table shows the correct conjugate acid and conjugate base for this ionisation?

	Conjugate acid	Conjugate base
A	NH_3	NH_4^+
B	NH_4^+	NH_3
C	NH_2^-	NH_4^+
D	NH_4^+	NH_2^-

17. The activation energies for the reactions

(1) $H_2(g) + I_2(g) \rightarrow 2HI(g)$

(2) $2HI(g) \rightarrow H_2(g) + I_2(g)$

are 165 kJ and 179 kJ respectively. The enthalpy change for reaction (2) is

A   −14 kJ

B   +14 kJ

C   −344 kJ

D   +344 kJ.

18. The standard enthalpy of formation of strontium chloride is the enthalpy change for which of the following reactions?

A $Sr(g) + Cl_2(g) \rightarrow SrCl_2(s)$

B $Sr(s) + Cl_2(g) \rightarrow SrCl_2(s)$

C $Sr^{2+}(g) + 2Cl^-(g) \rightarrow SrCl_2(s)$

D $Sr^{2+}(aq) + 2Cl^-(aq) \rightarrow SrCl_2(s)$

19. Consider the following thermochemical cycle which is not drawn to scale.

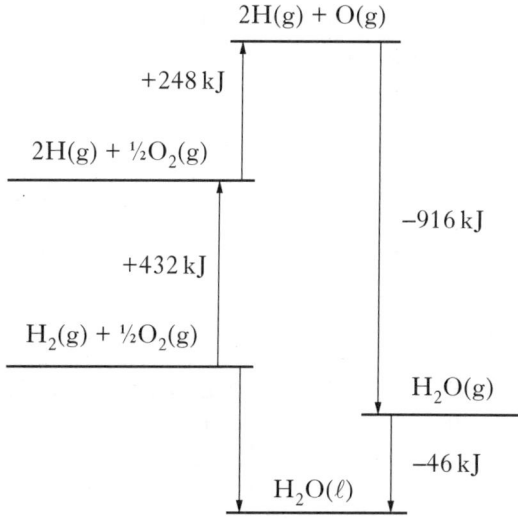

The enthalpy change for the reaction

$2H_2(g) + O_2(g) \rightarrow 2H_2O(\ell)$ is

A   −564 kJ

B   −282 kJ

C   +564 kJ

D   +1642 kJ.

20. In the presence of bright light, hydrogen and chlorine react explosively. One step in the reaction is shown below.

$$H_2(g) + Cl(g) \rightarrow HCl(g) + H(g)$$

The enthalpy change for this step can be represented as the bond enthalpy of

A　$(H-H) + (Cl-Cl)$

B　$(H-H) - (Cl-Cl)$

C　$(H-H) + (H-Cl)$

D　$(H-H) - (H-Cl)$.

21. The standard enthalpy of atomisation of bromine is the enthalpy change for the reaction

A　$\frac{1}{2}Br_2(s) \rightarrow Br(g)$

B　$\frac{1}{2}Br_2(\ell) \rightarrow Br(g)$

C　$\frac{1}{2}Br_2(g) \rightarrow Br(g)$

D　$Br_2(g) \rightarrow 2Br(g)$.

22. The enthalpy of solution of a compound can be calculated from its lattice enthalpy and the hydration enthalpies of its ions.

Using information from the Data Booklet, the correct value for enthalpy of solution of calcium chloride, in kJ mol^{-1}, is

A　-155

B　$+155$

C　-209

D　$+209$.

23. Which of the following reactions would show the greatest decrease in entropy?

A　$H_2(g) + F_2(g) \rightarrow 2HF(g)$

B　$KNO_3(s) \rightarrow KNO_2(s) + \frac{1}{2}O_2(g)$

C　$CO_3^{2-}(aq) + 2H^+(aq) \rightarrow H_2O(\ell) + CO_2(g)$

D　$CO_3^{2-}(aq) + CO_2(g) + H_2O(\ell) \rightarrow 2HCO_3^-(aq)$

24. Which of the following alcohols would have the greatest entropy at 90 °C?

A　Propan-1-ol

B　Propan-2-ol

C　Butan-1-ol

D　Butan-2-ol

25. Which of the following redox equations represents a reaction which is not feasible under standard conditions?

A　$F_2(g) + 2Cl^-(aq) \rightarrow 2F^-(aq) + Cl_2(g)$

B　$Cl_2(g) + 2Br^-(aq) \rightarrow 2Cl^-(aq) + Br_2(\ell)$

C　$F_2(g) + 2Br^-(aq) \rightarrow 2F^-(aq) + Br_2(\ell)$

D　$I_2(s) + 2Br^-(aq) \rightarrow 2I^-(aq) + Br_2(\ell)$

26. Propene can be produced by heating 1-bromopropane with ethanolic potassium hydroxide.

This reaction is an example of

A　reduction

B　hydrolysis

C　elimination

D　condensation.

27. The structures of three alcohols, **P**, **Q**, and **R** are shown.

P
$$H-\underset{\underset{H}{|}}{\overset{\overset{H}{|}}{C}}-\underset{\underset{H}{|}}{\overset{\overset{H}{|}}{C}}-\underset{\underset{H}{|}}{\overset{\overset{OH}{|}}{C}}-H$$

Q
$$H-\underset{\underset{H}{|}}{\overset{\overset{H}{|}}{C}}-\underset{\underset{H}{|}}{\overset{\overset{OH}{|}}{C}}-\underset{\underset{H}{|}}{\overset{\overset{OH}{|}}{C}}-H$$

R
$$H-\underset{\underset{H}{|}}{\overset{\overset{OH}{|}}{C}}-\underset{\underset{H}{|}}{\overset{\overset{OH}{|}}{C}}-\underset{\underset{H}{|}}{\overset{\overset{OH}{|}}{C}}-H$$

Which line in the table describes correctly the trends in boiling points and viscosities on moving from **P** to **Q** to **R**?

	Boiling point	Viscosity
A	increases	increases
B	increases	decreases
C	decreases	increases
D	decreases	decreases

28. Which of the following best describes the bonding in ethane?

A sp^2 hybridisation of the carbon atoms giving sigma bonds only

B sp^2 hybridisation of the carbon atoms giving sigma and pi bonds

C sp^3 hybridisation of the carbon atoms giving sigma bonds only

D sp^3 hybridisation of the carbon atoms giving sigma and pi bonds

29. Part of a possible chain reaction mechanism for chlorine reacting with methane is:

$$Cl_2 \rightarrow 2Cl\bullet$$
$$Cl\bullet + CH_4 \rightarrow HCl + CH_3\bullet$$
$$CH_3\bullet + Cl_2 \rightarrow CH_3Cl + Cl\bullet$$

Which of the following will **not** be a termination step in this reaction?

A $H\bullet + Cl\bullet \rightarrow HCl$

B $Cl\bullet + Cl\bullet \rightarrow Cl_2$

C $CH_3\bullet + CH_3\bullet \rightarrow C_2H_6$

D $CH_3\bullet + Cl\bullet \rightarrow CH_3Cl$

30. Pyridine, C_5H_5N, has the following structure:

Which line in the table shows the correct numbers of σ and π bonds in a molecule of pyridine?

	Number of σ bonds	Number of π bonds
A	3	11
B	6	3
C	11	3
D	12	3

31. The major product in the reaction of HCl with 2-methylpent-2-ene,

$$H_3C-\underset{}{\overset{\overset{CH_3}{|}}{C}}=CH-CH_2-CH_3 \text{ is}$$

A   2-chloro-2-methylpentane

B   3-chloro-2-methylpentane

C   2,3-dichloro-2-methylpentane

D   4-chloro-4-methylpentane.

[Turn over

32. A compound, **X**, reacts with the product of its own oxidation to form an ester.

    X could be

    A   propanal

    B   propan-1-ol

    C   propan-2-ol

    D   propanoic acid.

33. Which of the following amines does **not** have hydrogen bonds between its molecules in the liquid state?

    A $CH_3CH_2CH_2CH_2NH_2$

    B $CH_3CH_2NHCH_2CH_3$

    C $(CH_3)_2CHCH_2NH_2$

    D $(CH_3)_2NCH_2CH_3$

34. 1 mole of which of the following compounds would react with the largest volume of $1\ mol\ l^{-1}$ hydrochloric acid?

    A CH_3NHCH_3

    B $H_2NCH_2NH_2$

    C $CH_2OHCHOHCH_2OH$

    D   HO—⬡—NH_2

35. The conversion of benzene to monochlorobenzene using $Cl_2/FeCl_3$ involves

    A   nucleophilic addition

    B   nucleophilic substitution

    C   electrophilic addition

    D   electrophilic substitution.

36.

Which species initially attacks the benzene molecule in the above reaction?

    A NO_3^-

    B NO_2^+

    C HSO_4^-

    D NO_2

37. Which of the following is the geometric isomer of *trans*-1,2-dibromopropene?

A

B

C

D

38. The mass spectrum of an organic compound, empirical formula C_2H_4O, shows a peak for the parent ion at mass/charge ratio of 88.

    The organic compound could **not** be

    A   ethanal

    B   butanoic acid

    C   ethyl ethanoate

    D   methyl propanoate.

39. From which region of the electromagnetic spectrum is energy absorbed in the production of proton nmr spectra?

A   X-rays

B   Visible

C   Infra-red

D   Radio waves

40. A compound, which has molecular formula C_4H_8O, has only 2 peaks in its low resolution proton nmr spectrum.

A possible structural formula for this compound is

A $CH_3CH_2\underset{\underset{O}{\|}}{C}CH_3$

B $CH_3CH_2CH_2CHO$

C $CH_3-\underset{\underset{OH}{|}}{\overset{\overset{CH_3}{|}}{C}}-CH_3$

D $\begin{array}{c} H_2C-CH_2 \\ / \qquad \backslash \\ H_2C \qquad CH_2 \\ \backslash_{\,O\,}/ \end{array}$

[END OF SECTION A]

Candidates are reminded that the answer sheet for Section A MUST be placed INSIDE the front cover of your answer book.

[Turn over for SECTION B on *Page ten*

SECTION B   *Marks*

60 marks are available in this section of the paper.

All answers must be written clearly and legibly in ink.

1.   The compound, $Sn_2Ba_2(Sr_{0.5}Y_{0.5})Cu_3O_8$, has zero electrical resistance at 85 K.

    (*a*)   What name is given to this phenomenon?   **1**

    (*b*)   Which liquid coolant can be used economically and safely at this temperature?   **1**

        (2)

2.   When hydrogen is subjected to a high voltage in a gas discharge tube and the emitted light is passed through a prism the atomic emission spectrum produced is as shown below.

hydrogen emission spectrum (visible region)

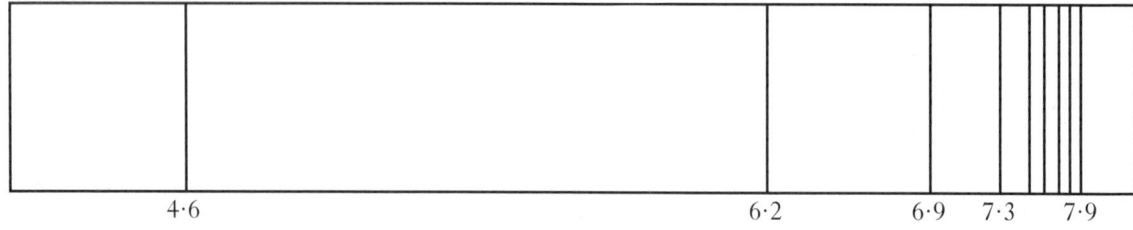

Frequency $v/10^{14}$ Hz

    (*a*)   Which line in the spectrum is red?   **1**

    (*b*)   The ionisation energy of hydrogen has a value of 1311 kJ mol^{-1}.

        (i)   Write the equation for the ionisation energy of hydrogen.   **1**

        (ii)   Calculate the wavelength of the light corresponding to this ionisation energy.   **3**

        (5)

Marks

3. When a mixture of nitrogen monoxide and nitrogen dioxide is cooled to −20 °C they react to form the clear blue liquid, dinitrogen trioxide.

$$NO + NO_2 \rightarrow N_2O_3$$

(a) The oxidation state of nitrogen is **different** in each of these three compounds.

Calculate the oxidation states of the nitrogen in NO and NO_2 respectively.   1

(b) Dinitrogen trioxide neutralises aqueous sodium hydroxide forming sodium nitrite and water.

The nitrite ion, NO_2^-, can be represented by two resonance structures.

One of these is

Draw the other resonance structure.   1

(c) In aqueous solution the nitrite ion can be oxidised to the nitrate ion.

Write the ion-electron equation for this oxidation.   1

(3)

[Turn over

Marks

4. Iron and manganese are transition metals which have many uses in industry.

   The electronic configuration for iron, in its ground state, is

   $$1s^2 2s^2 2p^6 3s^2 3p^6 3d^6 4s^2$$

   (a) In terms of s, p and d orbitals write down the electronic configurations of

       (i) Fe^{3+}

       (ii) Mn^{3+}

   in their ground states.   **2**

       (iii) Explain why the Fe^{3+} ion is more stable than the Mn^{3+} ion.   **1**

   (b) The transition metal titanium is the seventh most abundant element in the Earth's crust.

   Two of the reactions involved in the conversion of the ore ilmenite, $FeTiO_3$, into metallic titanium are shown below.

   Step 1—Ilmenite is reacted with concentrated sulphuric acid.

   $$FeTiO_3(s) + 3H_2SO_4(\ell) \rightarrow FeSO_4(aq) + Ti(SO_4)_2(aq) + 3H_2O(\ell)$$

   Step 2—After separation the titanium sulphate is reacted with sodium hydroxide.

   $$Ti(SO_4)_2(aq) + 4NaOH(aq) \rightarrow TiO_2(s) + 2H_2O(\ell) + 2Na_2SO_4(aq)$$

   How many kilograms of titanium oxide can theoretically be produced from 3·25 kg of ilmenite?   **2**

   (c) Transition metals can form a wide variety of complexes. One such complex is ammonium tetrachlorocuprate(II).

   Write the formula for this complex.   **1**

       (6)

Marks

5. The PPA "Complexometric Determination of Nickel using EDTA" has two main stages.

    Stage 1   Preparation of nickel(II) sulphate solution.

    Stage 2   Titration of the nickel(II) sulphate solution with EDTA.

The instructions for **Stage 1** are shown below.

1. Accurately weigh out approximately $2 \cdot 6\,g$ of hydrated nickel(II) sulphate, $NiSO_4.6H_2O$.

2. Transfer the hydrated nickel salt to a $100\,cm^3$ beaker, add $25\,cm^3$ of deionised water and stir to dissolve the solid.

3. Transfer the solution to a $100\,cm^3$ standard flask.

4.

5.

6. Stopper the flask and invert it several times to ensure the contents are thoroughly mixed.

(*a*) Complete the instructions for steps 4 and 5.   1

(*b*) In **Stage 2**, $25 \cdot 0\,cm^3$ of the nickel(II) sulphate solution were titrated against $0 \cdot 110\ mol\,l^{-1}$ EDTA solution.

The results of the titrations are shown below

	Rough titre	**1st titre**	**2nd titre**
Initial burette reading/cm^3	$2 \cdot 00$	$25 \cdot 90$	$10 \cdot 00$
Final burette reading/cm^3	$25 \cdot 90$	$49 \cdot 40$	$33 \cdot 60$
Volume of EDTA added/cm^3	$23 \cdot 90$	$23 \cdot 50$	$23 \cdot 60$

The equation for the reaction is represented by

$$Ni^{2+}(aq) + [EDTA]^{4-}(aq) \longrightarrow Ni[EDTA]^{2-}(aq)$$

   (i) Name the indicator used to detect the end-point of the titration in this PPA.   1

  (ii) EDTA acts as a hexadentate ligand. What shape is the complex ion $Ni[EDTA]^{2-}$?   1

 (iii) The accurate mass of the nickel(II) sulphate used was $2 \cdot 656\,g$.

        Calculate the percentage by mass of nickel present in the hydrated salt from these experimental results.   3

                                                     (6)

[Turn over

Marks

6. The standard free energy change for a chemical reaction is given by the expression

$$\Delta G^{\circ} = \Delta H^{\circ} - T\Delta S^{\circ}$$

The expression can be rearranged to give

$$\Delta G^{\circ} = -\Delta S^{\circ}T + \Delta H^{\circ}$$

Plotting values of ΔG° against T will therefore produce a straight line with gradient equal to $-\Delta S^{\circ}$.

The graph shows how ΔG° varies with temperature for a particular chemical reaction.

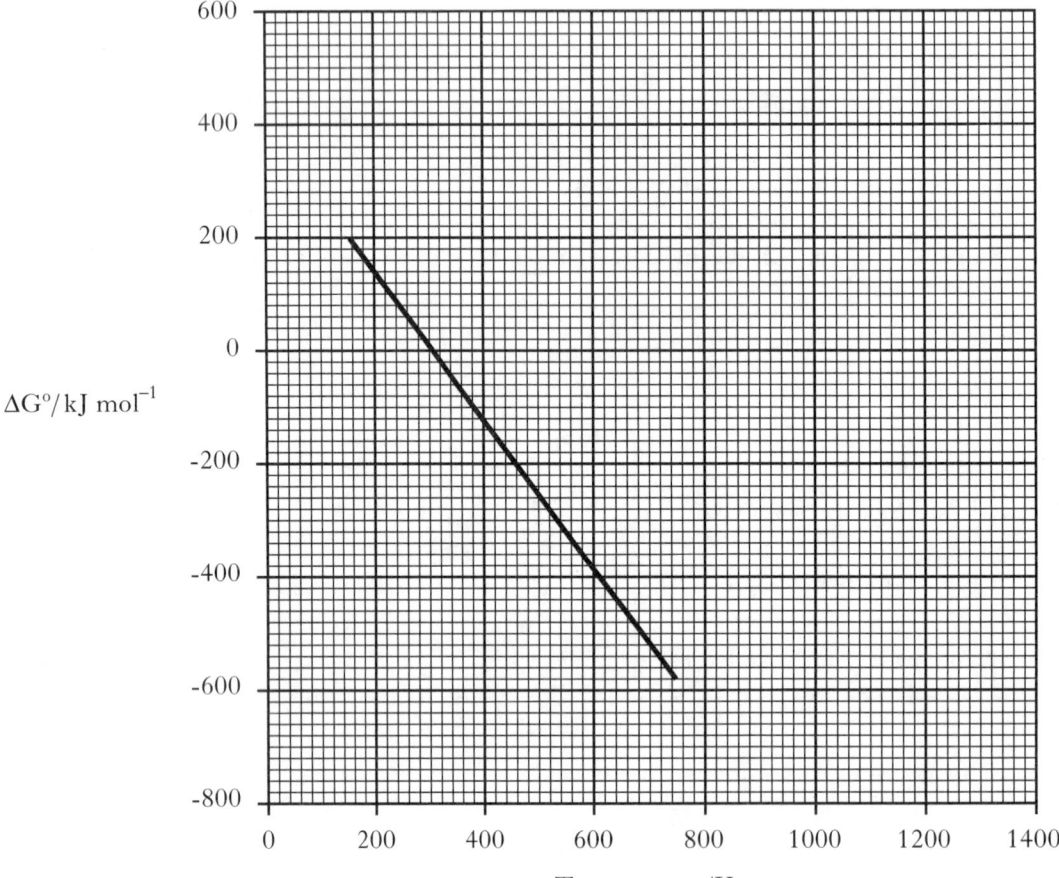

Use the graph to

(a) deduce the temperature at which the reaction just becomes feasible under standard conditions   1

(b) estimate the value of ΔH°, in kJ mol^{-1}, for the reaction   1

(c) calculate the value of ΔS°, in J K^{-1} mol^{-1}.   2

(4)

Marks

7. Consider the three reactions and their rate equations

   Reaction ① $2N_2O_5 \longrightarrow 4NO_2 + O_2$ Rate = $k[N_2O_5]$

   Reaction ② $2NO + Cl_2 \longrightarrow 2NOCl$ Rate = $k[NO]^2[Cl_2]$

   Reaction ③ $2NH_3 \longrightarrow N_2 + 3H_2$ Rate = $k[NH_3]^0$

   (a) What is the overall order of Reaction ②?   1

   (b) The graph below was plotted using experimental results from one of the reactions.

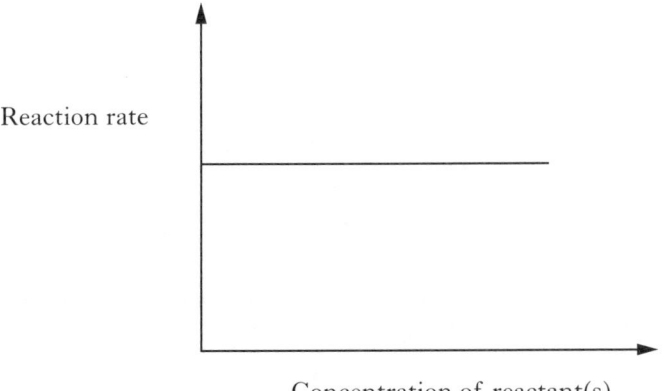

   Explain which of the reactions would give this graph.   1

   (c) For Reaction ②, when the concentrations of NO and Cl_2 are both 0.250 mol l^{-1}, the initial reaction rate is 1.43×10^{-6} mol l^{-1} s^{-1}.

   Use this information to calculate the rate constant, k, including the appropriate units.   2

                                                           (4)

8. The reaction between hydrogen peroxide and potassium bromide is used to generate bromine to disinfect water supplies.

   The ion-electron equations involved in this reaction are

          $Br_2(\ell) + 2e^- \longrightarrow 2Br^-(aq)$ $E^\circ = 1.07$ V

          $H_2O_2(aq) + 2H^+(aq) + 2e^- \longrightarrow 2H_2O(\ell)$ $E^\circ = 1.77$ V

   (a) Write the redox equation for the reaction.   1

   (b) Calculate the standard free energy change, in kJ mol^{-1}, for this reaction.   3

                                                           (4)

[Turn over

Marks

9. Buffer solutions are important in human biochemistry.

   (a) What is meant by a "buffer solution"? **1**

   (b) Suggest the name of a salt which could be mixed with propanoic acid to prepare an acidic buffer solution. **1**

   (c) The pH of an alkaline buffer solution can be found using the formula

   $$pH = pK_w - pK_b + \log \frac{[base]}{[salt]}$$

   where K_w is the ionic product of water

   and K_b is the dissociation constant of the base.

   $1·05\,g$ of ammonium nitrate, NH_4NO_3, is dissolved in $100\,cm^3$ of a $0·15\,mol\,l^{-1}$ ammonia solution at $25\,°C$.

   Calculate the pH of this buffer solution given that the pK_b for ammonia is $4·76$. **3**

   (5)

10. Chemists are developing compounds which block the ability of certain bacteria to bind to the surface of cells. This will help stop the spread of infection.

    (a) What name is given to the structural fragment of this type of compound which binds to a receptor? **1**

    (b) The diagram shows the structure of four of these compounds.

    (1)   (2)   (3)   (4)

    Draw the structural fragment which is common to these compounds which allows them to bind to the relevant receptor. **1**

    (2)

Marks

11. Meldrum's acid is a chemical named after the Scotsman, Andrew N. Meldrum who was the first to produce it.

Microanalysis showed that Meldrum's acid has a composition, by mass, of 50% C, 5·6% H, 44·4% O.

(*a*) Use the percentage composition to calculate the empirical formula of Meldrum's acid.

(Working must be shown) 1

(*b*) Meldrum initially thought the structure was

Compound A

The structure was later shown to be the isomer of **A** shown below.

Compound B

(i) What is the molecular formula of **A** and **B**? 1

(ii) The infra-red spectrum of isomer **A** would show a strong absorbance not shown by isomer **B**.

Identify the wave number range, in cm^{-1}, where this absorbance occurs. 1

(3)

[Turn over

Marks

12. Cinnamaldehyde is an aromatic compound found in cinnamon. It can also be prepared by the reaction of benzaldehyde and ethanal.

$$C_6H_5CHO + CH_3CHO \rightarrow C_6H_5CHCHCHO$$

(a) What type of reaction is this?   1

(b) Draw a full structural formula for cinnamaldehyde.   1

(c) All three of the carbonyl compounds shown above react with 2,4-dinitrophenylhydrazine, (Brady's reagent), forming solid derivatives.

The structure of 2,4-dinitrophenylhydrazine is

(i) Draw a structural formula of the compound formed when ethanal reacts with 2,4-dinitrophenylhydrazine.   1

(ii) The compound formed is impure.

How would this compound be purified?   1

(iii) How would the purified compound be used to show that the original carbonyl compound was ethanal?   1

(iv) 2,4-Dinitrophenylhydrazone derivatives have distinctive colours.

What colour is the 2,4-dinitrophenylhydrazone derivative of propanone?   1

    (6)

13. When sodium hydroxide solution was added to 2-bromomethylpropane an S_N1 reaction took place producing methylpropan-2-ol and hydrobromic acid.

(a) (i) What is meant by an S_N1 reaction?   2

(ii) Draw the structure of the carbocation intermediate formed in this reaction.   1

(b) Chloromethane reacts with sodium ethoxide in an S_N2 reaction.

(i) How is sodium ethoxide prepared in the laboratory?   1

(ii) Name the organic product of this S_N2 reaction.   1

    (5)

Marks

14. The structure of lactic acid is

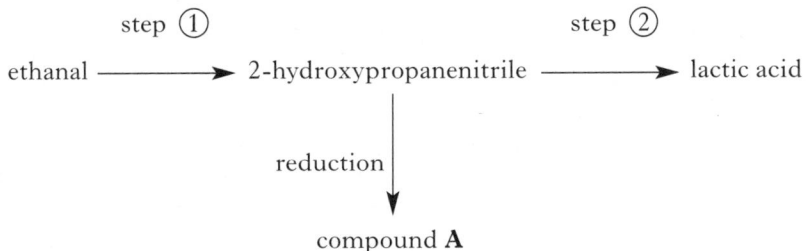

(a) What is the systematic name of lactic acid? 1

(b) Lactic acid contains an asymmetric carbon atom.

Identify, and **explain**, which one of the numbered carbon atoms is asymmetric. 1

(c) Lactic acid can be produced from ethanal by the reaction sequence below.

    step ①   step ②
ethanal ─────────→ 2-hydroxypropanenitrile ─────────→ lactic acid
                            │
                  reduction │
                            ↓
                     compound **A**

   (i) Which reagent could be used in step ① ? 1

   (ii) What type of reaction takes place in step ② ? 1

   (iii) Draw a structure for compound **A**. 1

(5)

[END OF QUESTION PAPER]

[BLANK PAGE]

ADVANCED HIGHER | ANSWER SECTION

CHEMISTRY ADVANCED HIGHER
2007

SECTION A

1. C	11. B	21. D	31. B
2. A	12. C	22. C	32. D
3. A	13. C	23. B	33. B
4. D	14. C	24. D	34. A
5. C	15. B	25. C	35. B
6. A	16. A	26. D	36. B
7. A	17. A	27. A	37. B
8. C	18. B	28. B	38. A
9. C	19. D	29. D	39. D
10. D	20. C	30. B	40. D

SECTION B

1. (a) +173·9 J K^{-1} mol^{-1}

   (b) 1529·6 K **or** 1530 K

2. (a) K = $\dfrac{[CH_4(g)][H_2S(g)]^2}{[CS_2(g)][H_2(g)]^4}$

   (b) 281

3. (a) proton donor/donates H$^+$ ions

   (b) HCO$_3^-$(aq)

   (c) 3·7

4. (a) (i) 1/1st order
      (ii) 1/1st order

   (b) Rate = k[H$_2$O$_2$][HI]

   (c) 0·0328 mol^{-1} l s^{-1}

5. (a) -181·4 kJ mol^{-1}

   (b) The concentration of both solutions should be 1 mol l^{-1}

   (c) 1·10 V

6. (a) Binds less strongly with metal ions than EDTA/ changes colour when it binds with metal ions.

   (b) 20·96% (21%)

   (c) Impurities/sample might be damp

7. (a) (i) 1s^2 2s^2 2p^6 3s^2 3p^6 3d^{10} 4s^1
      (ii) 1s^2 2s^2 2p^6 3s^2 3p^6 3d^7

   (b) (i) Cu$^+$(g) → Cu^{2+}(g) + e$^-$
      (ii) The electron in copper is being removed from a <u>full d subshell</u>/
          or Special stability associated with d^{10}.

8. (a) (i) Covalent (or polar covalent) in aluminium chloride
          Ionic in magnesium chloride.
      (ii) Al$_2$Cl$_6$

   (b) HCl   **or**   hydrogen chloride

(c)

(d) (i) Different ionic radii of positive ions/
         Ratios of ionic radii are different/
         Cs$^+$ ions bigger than Na$^+$ ions
    (ii) NaCl since the ionic radius ratio of
         Fe:O is similar to that of Na:Cl

         6:6 since the ionic radius ratio of
         Fe:O is similar to that of Na:Cl.

9. (a) Lone pairs/electrons available to form dative covalent bonds.

   (b) (i) III or 3 or 3+ or +3
      (ii) Oxidising agent/to oxidize Cr(III)/ electron acceptor

   (c) Hexacyanochromate(II)

10. (a) (i) Esters
       (ii) Solvents, perfumes

    (b) (i) 3-methylbutan-1-ol
       (ii) Condensation/esterification

11. (a) 4 and 5

    (b) Structural fragment which gives it pharmacological/biological activity **or**

        Structural fragment/section of a molecule that binds to receptor/triggers response

    (c) Agonist

12. (a) (i)   Hydrolysis
       (ii) Hydrochloric acid/HCl/concentrated HCl
       (iii) Recrystallisation
       (iv) 7·03 g/7·02 g

    (b) (i) Dehydration/elimination/cracking
       (ii)

13. (a) (i) 0·96 g C
             0·16 g H
             therefore 0·64 g O

       (ii)

C	H	O
$\dfrac{0.96}{12}$	$\dfrac{0.16}{1}$	$\dfrac{0.64}{16}$
$\dfrac{0.08}{0.04}$	$\dfrac{0.16}{0.04}$	$\dfrac{0.04}{0.04}$
2	4	1

C$_2$H$_4$O

(b) $C_4H_8O_2$

(c) (i)

$$H_3C-CH_2-CH_2-\overset{\displaystyle O}{\overset{\|}{C}}-O-H$$

Butanoic acid

(ii) Standard <u>reference</u> substance/for comparison purposes

CHEMISTRY ADVANCED HIGHER 2008

SECTION A

1. D	11. B	21. D	31. D
2. D	12. D	22. B	32. B
3. B	13. D	23. A	33. D
4. C	14. C	24. A	34. D
5. B	15. B	25. C	35. A
6. A	16. C	26. D	36. B
7. D	17. A	27. B	37. A
8. C	18. A	28. A	38. C
9. A	19. B	29. C	39. D
10. C	20. A	30. C	40. C

SECTION B

1. *(a)* Octahedral/square bipyramidal **or**
   5 bonded pairs and 1 lone pair

   (b) NF_3 has 4 electron pairs / BF_3 has 3 electron pairs
   NF_3 has tetrahedral arrangement of electron pairs / BF_3 has trigonal planar arrangement of electron pairs
   NF_3 has an extra electron pair / N has an extra electron pair

2. *(a)* Green

   (b) (i) $+3$ or $3+$ or 3 or III or three
   (ii) Tetraamminedichlorocobalt(III)
   (iii) $1s^2 2s^2 2p^6 3s^2 3p^6 3d^6$

3. *(a)* (Excited) electrons emitting energy as they fall back down to lower energy levels

   (b) Wavelength of emitted light is outwith the visible part of the spectrum
   Emission is in U.V. or I.R.
   Flame not hot enough

   (c) $178 \cdot 45$ kJ mol^{-1}

4. *(a)* (i) It coordinates through 2 sites to the platinum
   Forms 2 bonds to the Pt
   Pt is attached to 2 parts of the DNA
   Forms 2 dative bonds to the Pt
   (ii) Lone pairs of electrons
   Non bonded pairs of electrons
   Unbonded pairs of electrons

   (b)

   $$\begin{array}{ccc} Cl & & NH_3 \\ & \diagdown Pt \diagup & \\ H_3N & \diagup \quad \diagdown & Cl \end{array}$$

5. *(a)* (i) -478 kJ mol^{-1}
   (ii) 217 J K^{-1} mol^{-1}
   (iii) $-542 \cdot 7$ kJ mol^{-1} **or** -543 kJ mol^{-1}

   (b) -570 kJ mol^{-1}

6. *(a)* $1 \cdot 025 \times 10^{-3}$ **or** $1 \cdot 03 \times 10^{-3}$

   (b) $1 \cdot 025$ mol l^{-1}

7. *(a)* Step **X** $= 77$ kJ **or** 77 kJ mol^{-1}
   Step **Y** $= 382$ kJ **or** 382 kJ mol^{-1}

   (b) Lattice enthalpy
   Lattice formation enthalpy

   (c) $-535 \cdot 5$ kJ mol^{-1}

8. (a) Starch solution **or** starch **or** iotec

   (b) (i) 2.35×10^{-2} mol l^{-1}
       (ii) 2.625×10^{-2} mol l^{-1}

   (c) 0.894

   (d) (i) It decreases
       (ii) It would stay the same/No change/No effect

9. (a) $K_a = \dfrac{[H_3O^+(aq)] \times [F^-(aq)]}{[HF(aq)]}$

   (b) From the graph, $pK_a = 3.8$

       therefore $K_a = 1.58 \times 10^{-4}$

   (c) Sodium fluoride
       NaF **or** Na$^+$F$^-$ **or** F$^-$

   (d) Cresol red / alizarin red

10. (a) Elimination

    (b) (i) sp^2 hybridisation is a mixing of one s orbital and two
             p orbitals
        (ii) Sigma bonds - end on overlap of (atomic) orbitals
             Pi bonds - sideways overlap of (atomic) orbitals

    (c)

    H—C(H)(H)—C(H)(H)—C(H)(OH)—C(H)(H)—H

    (d) (i)   2nd order
        (ii) 1.32×10^{-4} l mol^{-1} s^{-1}
        (iii)

    OH$^-$ + [C$_3$H$_7$ / C(H)(H)—Br] →

    [HO----C(C$_3$H$_7$)(H)(H)----Br]$^{\ominus}$ → HO—C(C$_3$H$_7$)(H)—H

11. (a) (i) Sodium metal
             any reactive metal
             any Group 1 metal
             Ba
        (ii) (Hot) copper(II) oxide
             acidified dichromate
             acidified permanganate
             copper oxide
             acidified chromate

    (b) Ethoxypropane

    (c)

    CH$_3$CH$_2$—C(=O)—O—CH$_2$CH$_3$

12. (a) Condensation

    or addition + elimination

    (b) (i) By recrystallisation/crystallisation
        (ii) Measure melting point of derivative and compare with
             literature values/expected value/Data Book
             value/known value

(c) (i) Fehling's solution and blue to orange/brown with the
         isomer (propanal)
         Benedict's becomes orange/red with the isomer
         Tollens' reagent gives silver mirror with the isomer
         Acidified dichromate changes from orange to green
         with the isomer
         Acidified permanganate changes from purple to
         colourless with the isomer
         (Hot) copper(II) oxide changes from black to brown
         with the isomer
    (ii)

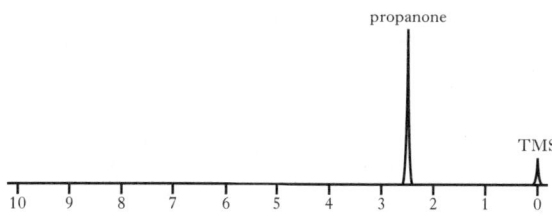

    (iii) A = CH$_3$$^+$
          B = CH$_3$CO$^+$ **or** C$_2$H$_3$O$^+$

13. (a) Electrophilic substitution

    (b) Br$_2$ and FeBr$_3$/FeCl$_3$/AlBr$_3$/AlCl$_3$

    (c) Sulphuric acid and nitric acid
        H$_2$SO$_4$ and HNO$_3$
        Concentrated/fuming H$_2$SO$_4$ + HNO$_3$

    (d) C$_6$H$_6$SO$_3$

    (e)

    [benzene ring]—CH$_2$Br

    [benzene ring]—CHBr$_2$

    [benzene ring]—CBr$_3$

CHEMISTRY ADVANCED HIGHER 2009

SECTION A

1.	D	21.	B
2.	D	22.	C
3.	A	23.	A
4.	C	24.	C
5.	A	25.	B
6.	C	26.	D
7.	D	27.	D
8.	B	28.	A
9.	D	29.	C
10.	C	30.	C
11	D	31.	B
12.	D	32.	C
13.	A	33.	C
14.	A	34.	D
15.	B	35.	B
16.	C	36.	A
17.	A	37.	B
18.	A	38.	B
19.	D	39.	C
20.	B	40.	C

SECTION B

1. (a) $1s^2\, 2s^2\, 2p^6\, 3s^2\, 3p^6$

   (b) (i) E= 78.3 (nm)

   (ii) $Ar(g) \rightarrow Ar^+(g) + e^-$

2. (a) $138\ J\ K^{-1}\ mol^{-1}$

   (b) $96\ kJ\ mol^{-1}$

   (c) = 696 K

3. (a) $Mg^{2+}(aq)$

   (b) Lattice (breaking) enthalpy

   (c) −728 kJ

   (d) $-322\ kJ\ mol^{-1}$

4. (a) Bond breaking H-H + $^1/_2$ (O=O)

   $= 432 + 248 \cdot 5 = 680 \cdot 5$
   Bond making 2 O-H = −916
   $\Delta H = (680 \cdot 5 - 916) = -235 \cdot 5\ (kJ\ mol^{-1})$

   (b) The above reaction has formed $H_2O(g)$ and more energy will be given out as it changes to $H_2O(l)$
   or
   Enthalpy of combustion forms $H_2O(l)$ at standard conditions but the above reaction has formed $H_2O(g)$

5. (a) $1\ mol\ l^{-1}\ H^+$ ions, 25°C (298K) and 1 atmosphere pressure

   (b) $2IO_3^- + 12H^+ + 10e^- \rightarrow I_2 + 6H_2O$

   (c) $-574 \cdot 2\ kJ\ mol^{-1}$

6. (a) (i) $HCOO^-$ **or** methanoate ion

   (ii) $K_a = \dfrac{[HCOO^-][H_3O^+]}{[HCOOH]}$

   (b) (i) $0 \cdot 0783\ mol\ l^{-1}$

   (ii) $2 \cdot 43$

7. (a) Rate = $k[CH_3COCH_3] [H^+]$

   (b) The H^+ is present at the **start** and the **end** of the reaction

   (c) (i) To quench/stop the reaction.
       To neutralise the acid.

       (ii) Starch solution **and** blue/black to colourless

8. (a) EDTA

   (b) nickel(II) ions are green **or** green/blue **or** coloured
       $Ni^{2+}(aq)$ absorb visible light

   (c) (i) it has lone pairs of electrons/non-bonding pairs of electrons

       (ii) 4

       (iii) gravimetric

       (iv) to prevent the complex from absorbing moisture
            or
            to allow the complex to cool in a dry atmosphere

9. (a)

   or

   $CH_3CH_2COCH_3$

   (b) (nucleophilic) substitution

   (c) Find its melting point and check with literature values

   (d)

   trans-but-2-ene   *cis*-but-2-ene

10. (a) H_2SO_4 and HNO_3
        or H_2SO_4 and $NaNO_3$

    (b) reduction

    (c) Ethanoic acid/CH_3COOH
        or ethanoyl chloride/CH_3COCl
        or ethanoic anhydride/$(CH_3CO)_2O$

11. (a) $C_4H_8O_2$

    (b) (i) carbonyl **or** C=O

        (ii) ester

    (c) $[CH_3CH_2CO]^+$
        $C_2H_5CO^+$

    (d) methyl propanoate

12. (a) +5 and +7

    (b) trigonal bipyramidal

    (c) sp^3d **or** sp^2d^2 **or** spd^3

    (d) Cl atom too small to accommodate 7 F atoms around it

13. (a)

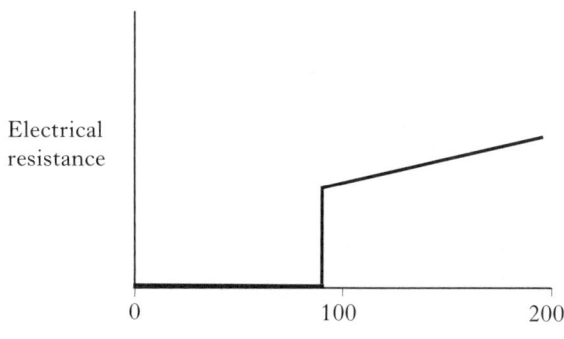

Electrical resistance vs Temperature/K (0, 100, 200)

(b) (i)

	Y	Ba	Cu	O
mass/g	13·4	41·2	28·6	16·8
moles	$\frac{13·4}{88·9}$	$\frac{41·2}{137·3}$	$\frac{28·6}{63·5}$	$\frac{16·8}{16·0}$
	= 0·151	= 0·300	= 0·450	= 1·05
mole ratio	1	2	3	7

(ii) +2·33

(iii) $Y_2Ba_4Cu_6O_{13}$ **or** $YBa_2Cu_3O_6$

CHEMISTRY ADVANCED HIGHER 2010

SECTION A

1.	D	21.	D
2.	C	22.	C
3.	D	23.	C
4.	B	24.	A
5.	B	25.	C
6.	D	26.	D
7.	A	27.	C
8.	B	28.	B
9.	D	29.	A
10.	C	30.	B
11	C	31.	D
12.	A	32.	B
13.	B	33.	A
14.	A	34.	B
15.	D	35.	D
16.	A	36.	A
17.	D	37.	C
18.	B	38.	B
19.	C	39.	A
20.	D	40.	C

SECTION B

1. (a) $E = \dfrac{Lhc}{\lambda}$ **or** $\dfrac{Lhc}{1000\lambda}$

$$E = \frac{6·02 \times 10^{23} \times 6·63 \times 10^{-34} \times 3·00 \times 10^8}{160 \times 10^{-9}}$$

$E = 748$ (kJ mol^{-1}) **or** 748361 J

(b) (i) 5
    (ii) trigonal bipyramidal

2. (a) +3 **or** III **or** 3

(b) tetraaquadichlorochromium(III)

(c)

3. (a) $\Delta H° = (-1676) - (-824) =$ **−852 kJ**

(b) $\Delta S° = [2(27) + 51] - [2(28) + 87] =$ **−38 JK^{-1}**

(c) $\Delta G° = \Delta H° - T\Delta S° = (-852) - 298(-38/1000)$
      $= -852 + 11·32 =$ **−841 kJ**

4. (a) (i) Brown fumes, fizzing, solution turning yellow, NO_2 forming
    (ii) Oxidising agent
    (iii) Max absorbance of permanganate/Green is the complementary colour of purple

(b) moles of Mn = $1.4 \times 10^{-4} \times 0.1$
   $= \mathbf{1.4 \times 10^{-5}}$
   mass of Mn = $1.4 \times 10^{-5} \times 54.9$
   $= \mathbf{7.686 \times 10^{-4}\ g}$
   % Mm = $(7.686 \times 10^{-4} / 0.19) \times 100$
   $= \mathbf{0.40\%}$

5. (a)

(b) b = 0
   c = −1 and d = −1

6. (a) (i) colourless to pink/colourless to purple
   (ii) $(16.5/1000) \times 0.02 \times 5/2 = \mathbf{0.000825\ moles}$
       (8.25×10^{-4})
   (iii) $0.000825 \times (1000/20) \times 88 = \mathbf{3.63g}$
   (iv) $4.49 - 3.63 - 0.06 = \mathbf{0.8g}$

(b)

K	H	C_2O_4
0.8/39	0.06/1	3.63/88
0.020	0.060	0.041

   X = 1   Y = 3   Z = 2

7. (a)
   $CH_3CH_2OH + CH_3COOH \rightleftharpoons CH_3COOCH_2CH_3 + H_2O$

(b) (i) No equilibrium in open system/System will not reach
       equilibrium
   (ii) At equilibrium:
       moles of water and ester = $0.70 - 0.24 = \mathbf{0.46}$
       moles of ethanoic acid = 0.24
       moles of ethanol = $0.68 - 0.46 = \mathbf{0.22}$
       K = [0.46][0.46]/[0.24][0.22] = **4.0**

8. (a)

(b) pKa = $-\log$Ka = $-\log 1.4 \times 10^{-5} = 4.85$
   pH = ½ pKa − ½ logc
   Substitute values $3.77 = 2.43 - ½ \log c$
   c = 0.0020mol l^{-1}

9. (a) Step two
   or
   $NO_2 + F \rightarrow NO_2F$

(b) $2NO_2 + F_2 \rightarrow 2NO_2F$
   or
   $NO_2 + ½F_2 \rightarrow NO_2F$

(c) 2nd order **or** 2

(d) k = Initial rate/$[NO_2][F_2]$ = **40 ℓ mol^{-1}s^{-1}**

10. (a) to give a higher yield
   or
   to reduce side reactions
   or
   to prevent charring

(b) sodium chloride (solution)/brine/salt water

(c) to dry the cyclohexene/dry the organic layer/drying
   agent/absorbs water/removes water

(d) Theoretical mass
   of cyclohexene $= \dfrac{82 \times 22.56}{100} = 18.5$ g
   % yield $= \dfrac{6.52 \times 100}{18.5} = 35\ \%$
   or
   Moles cyclohexanol = 22.56/100 = 0.2256 mol
   Moles cyclohexane = 6.52/82 = 0.0795 mol
   % yield $= \dfrac{0.0795}{0.2256} \times 100 = 35.2\ \%$ **or**   35%

11. (a) because but-2-ene has two different groups attached to
   each of the carbon atoms of the double bond
   or
   because in but-1-ene one of the carbon atoms of the
   double bond has identical groups (H) attached

(b)

(c) aqueous potassium (or sodium) hydroxide KOH(aq) or
   NaOH(aq) or LiOH(aq)
   or
   potassium (or sodium) hydroxide solution
   or
   aqueous alkali
   or
   alkali solution
   or
   water/H_2O

(d) aluminium chloride or $A\ell C\ell_3$
   or
   iron(III) chloride or $FeC\ell_3$
   or
   iron(III) bromide or $FeBr_3$
   or
   aluminium bromide or $A\ell Br_3$

(e)

12. (a) ethanal

(b) cyanohydrin

(c) hydrolysis/acid hydrolysis

(d) (i) van der Waals' forces
   (ii)

   Must be tetrahedral but dots and wedges can be
   replaced by solid lines

(iii) while one group would be able to bind to the appropriate region, the other two would not

or

the 3 'functional' groups fail to match the binding regions of the active site

or

only 1 group or 2 groups could bind (or match) the binding regions

or

The groups on the lactate ion no longer match the binding regions on the active site of the enzyme

or

The lactate ion no longer complements the binding region (of the active site)

or

The groups now fail to match the binding region (of the active site)

13. (a) Alcohols or alkanols

and

ethers

(b) (i)

    or $CH_3COCH_2CH_3$

(ii) butan-2-ol

CHEMISTRY ADVANCED HIGHER 2011

SECTION A

1.	A	21.	B
2.	B	22.	A
3.	A	23.	D
4.	C	24.	B
5.	C	25.	D
6.	C	26.	C
7.	A	27.	A
8.	D	28.	C
9.	C	29.	A
10.	D	30.	C
11	B	31.	A
12.	C	32.	B
13.	A	33.	D
14.	C	34.	B
15.	C	35.	D
16.	D	36.	B
17.	B	37.	C
18.	B	38.	A
19.	A	39.	D
20.	D	40.	D

SECTION B

1. (a) Superconductivity
Superconducting
Superconductor
Superconductance

(b) Liquid nitrogen/N_2

2. (a) The line at $4·6 \times 10^{14}$ Hz

(b) (i) $H(g) \rightarrow H^+(g) + e^-$
$H(g) - e^- \rightarrow H^+(g)$

(ii) $E = \dfrac{Lhc}{\lambda}$ **or** $E = \dfrac{Lhc}{1000\lambda}$

Wavelength, $\lambda = \dfrac{6·02 \times 10^{23} \times 6·63 \times 10^{-34} \times 3·00 \times 10^8}{1311000}$

$= 91·3 \times 10^{-9}$ m **or** $91·3$ nm **or** $9·13 \times 10^{-8}$m

or 91nm

3. (a) In NO, oxidation state is 2 or +2 or II **or** 2+
In NO_2, oxidation state is 4 or +4 or IV **or** 4+

(b)

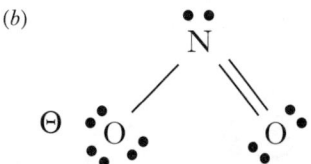

(c) $NO_2^-(aq) + H_2O(l) \rightarrow NO_3^-(aq) + 2H^+(aq) + 2e^-$

4. (a) (i) Fe^{3+} $1s^2 2s^2 2p^6 3s^2 3p^6 3d^5$
(ii) Mn^{3+} $1s^2 2s^2 2p^6 3s^2 3p^6 3d^4$
(iii) Fe^{3+} has half filled d-subshell
or All d-orbitals half filled in Fe^{3+}

(b) Moles of $FeTiO_3 = 3250/151·7 = $ **21·42**
Mass of $TiO_2 = $ n × FM $= 21·42 \times 79·9 = 1711$ g $= $ **1·71 kg**

(c) $(NH_4)_2[Cu(Cl)_4]$

5. (a) Step 4: Rinse beaker with deionised water, add washings to standard flask.
Step 5: Add deionised water up to mark on standard flask.

(b) (i) Murexide or ammonium purpurate
(ii) Octahedral

(iii) Average titre = 23·55 cm^3

No of moles of Ni in 100 cm^3 solution
= 0·02355 × 0·110 × 4 = 0·0104

% mass of Ni = $\dfrac{0·0104 × 58·7}{2·656}$ × 100 = **22·98%**

6. (a) T = 300 − 310 K

(b) ΔH° = 380 − 420 (kJ mol^{-1})

(c) Gradient of line = −1·3 (kJ K^{-1} mol^{-1})
or ΔS° = 1·22 to
1·40 kJ K^{-1} mol^{-1}

ΔS° = (+) 1220 to 1400 (J K^{-1} mol^{-1})

7. (a) Third order/3rd/3

(b) Reaction 3
Rate is independent of concentration of reactants

or rate is independent of concentration of ammonia

or Concentration of reactant has no effect on rate

(c) k = $\dfrac{\text{Rate}}{[NO]^2[Cl_2]}$ = $\dfrac{1·43 × 10^{-6}}{(0·250)^2(0·250)}$

= **9·15 × 10^{-5} l^2 mol^{-2} s^{-1}**

8. (a) $H_2O_2(aq) + 2H^+(aq) + 2Br^-(aq) \rightarrow Br_2(l) + 2H_2O(l)$

(b) ΔG° = −nFE°
= −2 × 96500 × 0·70
= −135·1 kJ mol^{-1}

9. (a) A solution in which the pH remains (approximately) constant when small amounts of acid, alkali or water are added, or a solution which resists pH changes when acid/alkali added

(b) Sodium propanoate or potassium propanoate

(c) [salt] = $\dfrac{0·15/80·0}{0·1}$ = 0·131 mol l^{-1}

pH = 14 − 4·76 + log $\dfrac{0·15}{0·131}$

pH = 14 − 4·76 + 0·059 = **9·30**

10. (a) Pharmacaphore

(b)

11. (a)

C	H	O
$\dfrac{50}{12}$	$\dfrac{5·6}{1}$	$\dfrac{44·4}{16}$
4·16	5·6	2·77
1·50	2·02	1

giving C$_3$H$_4$O$_2$

(b) (i) C$_6$H$_8$O$_4$

(ii) 2500 − 3500 (cm^{-1})
or 1700 − 1725 cm^{-1}

12. (a) Condensation

(b)

(c) (i)

(ii) Crystallisation/recrystallisation
(iii) Measure melting point **and** compare to known data/value
(iv) Yellow **or** orange **or** gold

13. (a) (i) Nucleophilic substitution/replacement by a nucleophile

First order **or** unimolecular

(ii)

(b) (i) Na in ethanol
(ii) Methoxyethane

14. (a) 2-hydroxypropanoic acid

(b) Carbon atom ② because it has 4 different groups attached

(c) (i) KCN or NaCN or HCN
or correct names
(ii) Hydrolysis/acid hydrolysis
(iii)

or CH$_2$CHOHCH$_2$NH$_2$

Hey! I've done it

Published by Bright Red Publishing Ltd, 6 Stafford Street, Edinburgh, EH3 7AU
Tel: 0131 220 5804, Fax: 0131 220 6710, enquiries: sales@brightredpublishing.co.uk,
www.brightredpublishing.co.uk

Official SQA answers to 978-1-84948-229-5
2007-2011